STUDENT SOLUTIONS MANUAL
to accompany

APPLIED CALCULUS
for
BUSINESS, SOCIAL SCIENCES,
and
LIFE SCIENCES

preliminary edition

DEBORAH HUGHES-HALLETT
Harvard University

ANDREW M. GLEASON
Harvard University

PATTI FRAZER LOCK
St. Lawrence University

DANIEL FLATH
University of South Alabama

et al.
•
prepared by

DEAN CHUNG • MEGAN McCALLION • REBECCA RAPOPORT • JENNIE YODER

This material is based upon work supported by the National Science Foundation under Grant No. DUE-9352905.

Copyright © 1996 by John Wiley & Sons, Inc.

All rights reserved.

Reproduction or translation of any part of this work
beyond that permitted by Sections 107 and 108 of
the 1976 United States Copyright Act without the
permission of the copyright owner is unlawful.
Requests for permission or further information
should be addressed to the Permissions Department,
John Wiley & Sons, Inc.

ISBN 0-471-11117-1

Printed in the United States of America

10 9 8 7 6 5 4 3 2

CONTENTS

CHAPTER 1 1

CHAPTER 2 29

CHAPTER 3 45

CHAPTER 4 55

CHAPTER 5 63

CHAPTER 6 79

CHAPTER 7 89

CHAPTER 8 101

CHAPTER 9 115

APPENDIX 129

CHAPTER 1

Answers for Section 1.1

1. **(I)** The first graph does not match any of the given stories. In this picture, the person keeps going away from home, but his speed decreases as time passes. So a story for this might be: *I started walking to school at a good pace, but since I stayed up all night studying calculus, I got more and more tired the farther I walked.*
   (II) This graph matches (b), the flat tire story. Note the long period of time during which the distance from home did not change (the horizontal part).
   (III) This one matches (c), in which the person started calmly but sped up.
   (IV) This one is (a), in which the person forgot her books and had to return home.

5. At first, as the number of workers increases, productivity also increases. As a result, the graph of the curve goes up initially. After a certain point the curve goes downward; in other words, as the number of workers increases beyond that point, productivity decreases. This might be due either to the inefficiency inherent in large organizations or simply to workers getting in each other's way as too many are crammed on the same line.

9. The x values for which the function is defined are all x between 0 and 10. Thus the domain is

$$0 \leq x \leq 10.$$

   Meanwhile, the possible y values are all values between roughly 1.6 (its minimum) and 5.5 (its maximum). Thus the range is approximately

$$1.6 \leq y \leq 5.5$$

13. (a) We know that the equilibrium point is the point where the supply and demand curves intersect. Looking at Fig 1.13 we see that the price at which they intersect is $10 per unit and the corresponding quantity is 3000 units.
    (b) We know that the supply curve climbs upwards while the demand curve slopes downwards. Thus we know that at the price of $12 per unit the supplier will be willing to produce 3500 units while the consumer will be ready to buy 2500 units. Thus we see that when the price is above the equilibrium point, more items will be produced than the consumer will be willing to buy. Thus the producer ends up wasting money by producing that which will not be bought, so he is better off lowering the price.
    (c) Looking at the point on the rising curve where the price is $8 per unit, we see that that the supplier will be willing to produce 2500 units, whereas looking at the point on the downwards sloping curve where the price is $8 per unit, we see that the consumer will be willing to buy 3500 units. Thus we see that when the price is less than the equilibrium price, the consumer is willing to buy more products than the supplier is making and the supplier can thus make more money by producing more units or raising the price.

17. (a) We know that as the price per unit increases the quantity in the supply curve increases while the quantity in the demand curve decreases. Thus we get that Table 1.2 is the demand curve (since as the price increases the quantity decreases) while Table 1.3 is the supply curve (since as the price increases the quantity increases.)
   (b) Looking at Table 1.2 - the demand curve data - we see that a price of $155 gives a quantity of roughly 14.
   (c) Looking at Table 1.3 - the supply curve data - we see that a price of $155 gives a quantity of roughly 24.
   (d) Since supply exceeds demand at a price of $155, the shift would be to a lower price.
   (e) We were asked for the price at which the consumers would buy at least 20 items. Looking at the data for the demand curve - Table 1.2 - we see that if the price is less than or equal to $143 the consumers would buy at least 20 items.
   (f) Looking at the data for the supply curve - Table 1.3 - we see that if the price is greater than or equal to $110 the supplier will produce at least 20 items.

21. Looking at the given data, it seems that Galileo's hypothesis was incorrect. The first table suggests that velocity is not a linear function of distance, since the increases in velocity for each foot of distance are themselves getting smaller. Moreover, the second table suggests that velocity is instead proportional to *time*, since for each second of time, the velocity increases by 32 ft/sec.

Answers for Section 1.2

1. (a) (V)
   (b) (IV)
   (c) (I)
   (d) (VI)
   (e) (II)
   (f) (III)

5. $y - c = m(x - a)$

9. (a) On the interval from 0 to 1 the value of y decreases by 2. On the interval from 1 to 2 the value of y decreases by 2. And on the interval from 2 to 3 the value of y decreases by 2. Thus the function has a constant rate of change and it is therefore linear.
   (b) On the interval from 15 to 20 the value of s increases by 10. On the interval from 20 to 25 the value of s increases by 10. And on the interval from 25 to 30 the value of s increases by 10. Thus the function has a constant rate of change and is linear.
   (c) On the interval from 1 to 2 the value of w increases by 5. On the interval from 2 to 3 the value of w increases by 8. Thus we see that the slope of the function is not constant and therefore the function is not linear.

13. (a) Rewriting this equation we get

$$3x + 4y = -12$$
$$4y = -12 - 3x$$
$$y = -3 - 0.75x$$

We know that when $x = 0$, we have $y = -3$. Thus the y-intercept is $(0, -3)$. We know that when $y = 0$ we have

$$y = -3 - 0.75x$$

$$0 = -3 - 0.75x$$
$$3 = -0.75x$$
$$x = -4$$

Thus the x-intercept is $(-4, 0)$.

(b) For the line $3x + 4y = -12$, the x-intercept is $(-4, 0)$ and the y-intercept is $(0, -3)$. The distance between these two points is

$$d = \sqrt{(-4-0)^2 + (0-(-3))^2} = \sqrt{16+9} = \sqrt{25} = 5.$$

17. (a) We know that the function will take on the form

$$f(t) = mt + b.$$

We know that the slope will be

$$\text{slope} = \frac{19.72 - 18.48}{0 - 1} = -1.24$$

We also know that when $t = 0$, we have $f(t) = 19.72$. Thus the vertical intercept is

$$b = 19.72$$

Hence we get

$$f(t) = -1.24t + 19.72$$

(b)

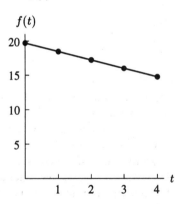

Figure 1.1

21. We know that a formula for passengers versus price will take the form

$$p = md + b$$

where p is the number of passengers on the boat when the price of a tour is d dollars. We know two points on the line thus we know that the slope is

$$\text{slope} = \frac{650 - 500}{20 - 25} = \frac{150}{-5} = -30.$$

Thus the function will look like

$$p = -30d + b.$$

Plugging in the point $(20, 650)$ we get

$$p = -30d + b$$
$$650 = (-30)(20) + b$$
$$= -600 + b$$
$$b = 1250$$

Thus a formula for the number of passengers as a function of tour price is

$$p = -30d + 1250.$$

25. (a) $R = k(350 - H)$, where k is a positive constant.
    If H is greater than $350°$, the rate is negative, indicating that a very hot yam will cool down toward the temperature of the oven.
    (b) Letting H_0 equal the initial temperature of the yam, the graph of R against H looks like:

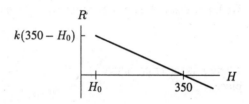

Note that by the temperature of the yam, we mean the average temperature of the yam, since the yam's surface will be hotter than its center.

Answers for Section 1.3

1. (a) We know that regardless of the number of rides one takes, one must pay $7 to get in. After that, for each ride you must pay another $1.50, thus the function $C(n)$ is

$$C(n) = 7 + 1.5n$$

   (b) Plugging in the values $n = 2$ and $n = 8$ into our formula for $C(n)$ we get

$$C(2) = 7 + 1.5(2) = 7 + 3 = \$10$$

   and

$$C(8) = 7 + 1.5(8) = 7 + 12 = \$19.$$

(c)

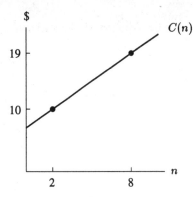

Figure 1.2

5. (a) We know that the cost function will be of the form

$$C(q) = m \cdot q + b$$

where m is the slope of the graph and b is the vertical intercept. We also know that the fixed cost is the vertical intercept and the variable cost is the slope. Thus we have

$$C(q) = 5000 + 30q.$$

We know that the revenue function will take on the form

$$R(q) = pq$$

where p is the price charged per unit. In our case the company sells the chairs at $50 a piece so

$$R(q) = 50q.$$

(b)

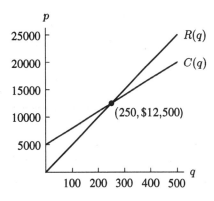

Figure 1.3

(c) We know that the break-even point is the number of chairs that the company has to sell so that the revenue will equal the cost of producing these chairs. In other words, we are looking for q such that

$$C(q) = R(q).$$

Solving we get
$$C(q) = R(q)$$
$$5000 + 30q = 50q$$
$$20q = 5000$$
$$q = 250$$

Thus the break-even point is 250 chairs.
Graphically, it is the point in Fig 1.3 where the cost function intersects the revenue function.

9. (a) We know that the fixed cost for this company is the amount of money it takes to produce zero units, or simply the vertical intercept of the graph. Thus the
$$\text{fixed cost} = \$1000.$$

(b) We know that the variable cost is the price the company has to pay for each additional unit, or in other words, the slope of the graph. We know that
$$C(0) = 1000$$
and looking at the graph we also see that
$$C(200) = 4000.$$
Thus the slope of the line, or the variable cost, is
$$\text{variable cost} = \frac{4000 - 1000}{200 - 0} = \frac{3000}{200} = \$15.$$

(c) $C(q)$ gives the price that the company will have to pay for the production of q units. Thus if
$$C(100) = 2500$$
we know that it will cost the company $2500 to produce 100 items.

13. (a) We know that the function for the value of the robot at time t will be of the form
$$V(t) = m \cdot t + b.$$
We know that at time $t = 0$ the value of the robot is $15,000. Thus the vertical intercept b is
$$b = 15,000.$$
We know that m is the slope of the line. Thus
$$m = \frac{0 - 15,000}{10 - 0} = \frac{-15,000}{10} = -1500.$$
Thus we get
$$V(t) = -1500t + 15,000.$$

(b) The value of the robot in three years is
$$V(3) = -1500(3) + 15,000 = -4500 + 15,000 = \$10,500.$$

17. (a) Let
$$I = \text{number of Indian peppers}$$
$$M = \text{number of Mexican peppers.}$$
Then (from the given information)
$$1{,}200I + 900M = 14{,}000$$
is the Scoville constraint.

(b) Solving for I yields
$$I = \frac{14{,}000 - 900M}{1{,}200}$$
$$= \frac{35}{3} - \frac{3}{4}M.$$

21. In this problem, the tax is imposed on the consumer and thus the perceived price that the consumer pays is $1.05p$. Thus, the equilibrium price occurs when the new demand function equals the supply function:
$$D(1.05p) = S(p)$$
$$100 - 2(1.05p) = 3p - 50$$
$$150 = 5.1p$$
$$p = 29.41$$

The equilibrium quantity q is then the supply value at the equilibrium price,
$$q = S(29.41) = 3(29.41) - 50 = 88.23 - 50 = 38.23$$

Then we can see that the producer pays as tax the old equilibrium price minus the new equilibrium price:
$$30 - 29.41 = 0.59$$

The consumer pays as tax the new (after tax) equilibrium price minus the old equilibrium price:
$$(1.05)p - 30 = (29.41)(1.05) - 30 = 30.88 - 30 = 0.88$$

In comparison to Problem 20, the taxes are slightly lower although not by much. Furthermore the consumer and the producer pay similar amounts in each case, their relative proportions remaining the same.

Answers for Section 1.4

1.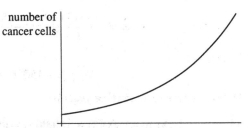

5. (a) This is the graph of a linear function, which increases at a constant rate, and thus corresponds to $k(t)$, which increases by 0.3 over each interval of 1.
   (b) This graph is concave down, so it corresponds to a function whose increases are getting smaller, as is the case with $h(t)$, whose increases are 10, 9, 8, 7, and 6.
   (c) This graph is concave up, so it corresponds to a function whose increases are getting bigger, as is the case with $g(t)$, whose increases are 1, 2, 3, 4, and 5.

9. Each increase of 1 in t seems to cause $g(t)$ to decrease by a factor of 0.8, so we expect an exponential function with base 0.8. To make our solution agree with the data at $t = 0$, we need a coefficient of 5.50, so our completed equation is

$$g(t) = 5.50(0.8)^t.$$

13. We look for an equation of the form $y = y_0 a^x$ since the graph looks exponential. The points $(0, 3)$ and $(2, 12)$ are on the graph, so

$$3 = y_0 a^0 = y_0$$

and

$$12 = y_0 \cdot a^2 = 3 \cdot a^2, \quad \text{giving} \quad a = \pm 2.$$

Since $a > 0$, our equation is $y = 3(2^x)$.

17. (a) The formula is $Q = Q_0 \left(\frac{1}{2}\right)^{(t/1620)}$.
    (b) The percentage left after 500 years is

$$\frac{Q_0 \left(\frac{1}{2}\right)^{(500/1620)}}{Q_0}.$$

The Q_0's cancel giving

$$\left(\frac{1}{2}\right)^{(500/1620)} \approx 0.807,$$

so 80.7% is left.

21. The doubling time t depends only on the growth rate; it is the solution to

$$2 = (1.02)^t,$$

since 1.02^t represents the factor by which the population has grown after time t. Trial and error shows that $(1.02)^{35} \approx 1.9999$ and $(1.02)^{36} \approx 2.0399$, so that the doubling time is about 35 years.

25. (a) Its cost in 1989 would be $1000 + \frac{1290}{100} \cdot 1000 = 13900$ cruzados.
    (b) The monthly inflation rate r solves

$$\left(1 + \frac{1290}{100}\right)^1 = \left(1 + \frac{r}{100}\right)^{12},$$

since we compound the monthly inflation 12 times to get the yearly inflation. Solving for r, we get $r = 24.52\%$. Notice that this is much different than $\frac{1290}{12} = 107.5\%$.

Answers for Section 1.5

1. $y = x^{2/3}$ is larger as $x \to \infty$.

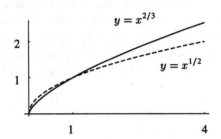

5. The odd powers are increasing everywhere, whereas the even powers are U-shaped. For $x > 0$, the highest powers are largest for big x, and the smallest are largest for x near 0. This is as expected from the graphs in the text.

   (a)

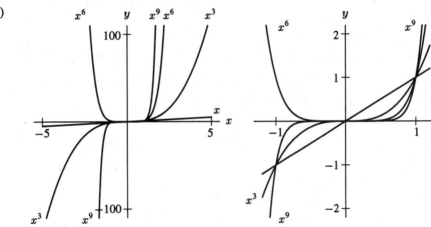

Figure 1.4

   (b)

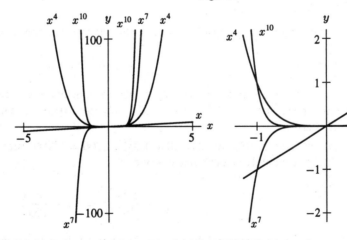

Figure 1.5

9. As $x \to \infty$, $f(x) = x^5$ has the largest positive values. As $x \to -\infty$, $g(x) = -x^3$ has the largest positive values.

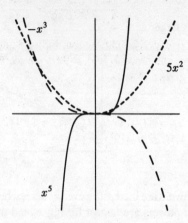

13. 3^x is always positive while x^3 is negative for $x < 0$, so we know that $3^x > x^3$ for $x < 0$. Looking at the large-scale graph, we see that for $x > 4$, 3^x is also clearly bigger than x^3. We zoom in on the interval $(0, 4)$ to see the behavior there, shown by the next graph. We see that the graphs approach very close to each other near $x = 3$ (where the values are equal), but elsewhere, $3^x > x^3$. To figure out what is going on near $x = 3$, we zoom in again, and notice that on the interval from about 2.5 to 3, $x^3 > 3^x$. Thus $3^x > x^3$ if $x < 2.5$ (approximately) or $x > 3$.

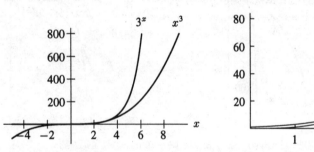
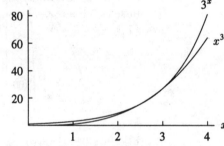

Figure 1.6   Figure 1.7

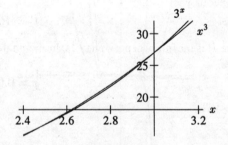

Figure 1.8

Alternatively, graph $f(x) = 3^x - x^3$ for $0 \le x \le 3.5$ and $-2 \le y \le 4$. This shows $3^x > x^3$ for $x > 3$ and $x < 2.48$.

17. Looking at g, we see that the ratio of the values is:

$$\frac{3.12}{3.74} \approx \frac{3.74}{4.49} \approx \frac{4.49}{5.39} \approx \frac{5.39}{6.47} \approx \frac{6.47}{7.76} \approx 0.83.$$

Thus g is an exponential function, and so f and k are the power functions. Each is of the form ax^2 or ax^3, and since $k(1.0) = 9.01$ we see that for k, the constant coefficient is 9.01. Trial and error gives

$$k(x) = 9.01x^2,$$

since $k(2.2) = 43.61 \approx 9.01(4.84) = 9.01(2.2)^2$. Thus $f(x) = ax^3$ and we find a by noting that $f(9) = 7.29 = a(9^3)$ so

$$a = \frac{7.29}{9^3} = 0.01$$

and $f(x) = 0.01x^3$.

Answers for Section 1.6

1. For $20 \leq x \leq 100$, $0 \leq y \leq 1.2$, this function looks like a horizontal line at $y = 1.0725\ldots$ (In fact, the graph approaches this line from below.) Now, $e^{0.07} \approx 1.0725$, which strongly suggests that, as we already know, as $x \to \infty$, $\left(1 + \frac{0.07}{x}\right)^x \to e^{0.07}$.

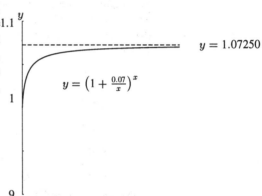

5. $e^{0.06} = 1.0618365$, so the effective annual rate $\approx 6.18365\%$.

9. We know that the formula for the balance after t years in an account which is compounded continuously is

$$B(t) = P \cdot e^{rt}$$

where P is the initial deposit and r is the nominal rate. In our case we are told that the nominal rate is

$$r = 0.08$$

so we have

$$B(t) = P \cdot e^{0.08t}.$$

We are asked to find the value for P such that after three years the balance would be $10,000. That is, we are asked for P such that

$$10{,}000 = P \cdot e^{0.08 \cdot 3}.$$

Solving we get

$$10,000 = P \cdot e^{0.08 \cdot 3}$$
$$= P \cdot e^{0.24}$$
$$\approx P(1.27125)$$
$$P = \frac{10,000}{1.27125}$$
$$= 7866.28$$

Thus, $7866.28 should be deposited into this account so that after three years the balance would be $10,000.

13. We use the formula $A = A_0(1 + \frac{r}{n})^{nt}$, where t is in years, r is annual rate of interest and n is the number of times interest is compounded per year.

(a) Compounding daily,

$$A = 450,000 \left(1 + \frac{0.06}{365}\right)^{(213)(365)}$$
$$= 450,000 \, (1.00016438)^{77745}$$
$$\approx \$1.59602561 \times 10^{11}$$

This amounts to approximately $160 billion.

(b) Compounding yearly,

$$A = 450,000 \, (1 + 0.06)^{213}$$
$$= 450,000(1.06)^{213} = 450,000(245555.29)$$
$$= \$1.10499882 \times 10^{11}$$

This is only $110.5 billion.

(c) We first wish to find the interest that will accrue during 1990. For 1990, the principal is 1.105×10^{11}. At 6% annual interest, during 1990 the money will earn

$$0.06 \times \$1.105 \times 10^{11} = \$6.63 \times 10^9.$$

The number of seconds in a year is

$$\left(365 \frac{\text{day}}{\text{year}}\right)\left(24 \frac{\text{hr}}{\text{day}}\right)\left(60 \frac{\text{min}}{\text{hr}}\right)\left(60 \frac{\text{sec}}{\text{min}}\right) = 31536000 \text{ sec}.$$

Thus, over 1990, interest is accumulating at the rate of

$$\frac{\$6.63 \times 10^9}{31536000 \text{ sec}} \approx \$210.24 \text{ /sec}.$$

Answers for Section 1.7

1. $y = \ln e^x$ is a straight line with slope 1, passing through the origin. This is so because $y = \ln e^x = x \ln e = x \cdot 1 = x$. So this function is really $y = x$ in disguise.

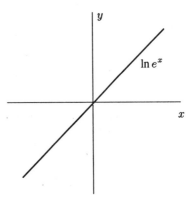

5. Taking natural logs of both sides we get
$$\ln 130 = \ln(10^t).$$
This gives
$$t \ln 10 = \ln 130$$
or in other words
$$t = \frac{\ln 130}{\ln 10} \approx 2.1139.$$

9. Dividing both sides by 10 we get
$$5 = 3^t.$$
Taking natural logs of both sides gives
$$\ln(3^t) = \ln 5.$$
This gives
$$t \ln 3 = \ln 5$$
or in other words
$$t = \frac{\ln 5}{\ln 3} \approx 1.465.$$

13. Taking natural logs of both sides we get
$$\ln(e^{3t}) = \ln 100.$$
This gives
$$3t = \ln 100$$
or in other words
$$t = \frac{\ln 100}{3} \approx 1.535.$$

17. Taking natural logs of both sides we get
$$\ln(7 \cdot 3^t) = \ln(5 \cdot 2^t)$$
which gives
$$\ln 7 + \ln(3^t) = \ln 5 + \ln(2^t)$$
or in other words
$$\ln 7 - \ln 5 = \ln(2^t) - \ln(3^t)$$
This gives
$$\ln 7 - \ln 5 = t \ln 2 - t \ln 3 = t(\ln 2 - \ln 3)$$
Thus we get
$$t = \frac{\ln 7 - \ln 5}{\ln 2 - \ln 3} \approx -0.8298.$$

21.
$$e^{\ln(x+1)} = x + 1$$

25. (a) Plugging in $t = 0.5$ in $P(t)$ gives
$$\begin{aligned} P(0.5) &= 1000 e^{-0.5 \cdot 0.5} \\ &= 1000 e^{-0.25} \\ &\approx 1000(0.7788) \\ &= 778.8 \end{aligned}$$
Plugging in $t = 1$ in $P(t)$ gives
$$\begin{aligned} P(1) &= 1000 e^{-0.5} \\ &\approx 1000(0.6065) \\ &= 606.5 \end{aligned}$$

(b) Plugging in $t = 3$ in $P(t)$ gives
$$\begin{aligned} P(3) &= 1000 e^{-0.5 \cdot 3} \\ &= 1000 e^{-1.5} \\ &\approx 1000(0.2231) \\ &= 223.1 \end{aligned}$$
This tells us that after 3 years the fish population has gone down to about 22% of the initial population.

(c) We are asked to find the value of t such that $P(t) = 100$. That is, we are asked to find the value of t such that
$$100 = 1000 e^{-0.5t}.$$
Solving this gives
$$\begin{aligned} 100 &= 1000 e^{-0.5t} \\ 0.1 &= e^{-0.5t} \\ \ln 0.1 &= \ln e^{-0.5t} = -0.5t \\ t &= \frac{\ln 0.1}{-0.5} \\ &\approx 4.6 \end{aligned}$$

Thus after roughly 4.6 years, the trout population will have decreased to 100.
(d) A graph of the population is given in Fig 1.9

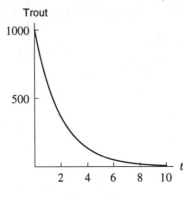

Figure 1.9

Looking at the graph, we see that as time goes on there are less and less trout in the pond. A possible cause for this may be the presence of predators in the pond, such as fisherman. The rate at which the fish die decreases due to the fact that there are fewer fish in the pond, which means that there are fewer fish that have to try to survive. Notice that the survival chances of any given fish remains the same, it is just the overall mortality that is decreasing.

29. Let $t =$ number of years since 1980. Then the number of vehicles, V, in millions, at time t is given by
$$V = 170(1.04)^t$$
and the number of people, P, in millions, at time t is given by
$$P = 227(1.01)^t.$$
There is an average of one vehicle per person when $\dfrac{V}{P} = 1$, or $V = P$. Thus, we must solve for t the equation:
$$170(1.04)^t = 227(1.01)^t,$$
which implies
$$\left(\frac{1.04}{1.01}\right)^t = \frac{(1.04)^t}{(1.01)^t} = \frac{227}{170}$$
Taking logs on both sides,
$$t \log \frac{1.04}{1.01} = \log \frac{227}{170}.$$
Therefore,
$$t = \frac{\log\left(\frac{227}{170}\right)}{\log\left(\frac{1.04}{1.01}\right)} \approx 9.9 \text{ years}.$$
So there was, according to this model, about one vehicle per person in 1990.

Answers for Section 1.8

1. We know that the formula for the balance in an account after t years is
$$B(t) = Pe^{rt}$$
where P is the initial deposit and r is the nominal rate. In our case the initial deposit is $12,000 that is
$$P = 12,000$$
and the nominal rate is
$$r = 0.08.$$
Thus we get
$$B(t) = 12,000e^{0.08t}.$$
We are asked to find t such that $B(t) = 20,000$, that is we are asked to solve
$$20,000 = 12,000e^{0.08t}$$
$$e^{0.08t} = \frac{20,000}{12,000} \approx 1.667$$
$$\ln e^{0.08t} = \ln 1.667$$
$$0.08t = \ln 1.667$$
$$t = \frac{\ln 1.667}{0.08} \approx 6.39$$
Thus after roughly 6.39 years there will be $20,000 in the account.

5. We know that a function modeling the quantity of substance at time t will be an exponential decay function satisfying the fact that at $t = 10$, 4% of the substance has decayed, or in other words 96% of the substance is left. Thus a formula for the quantity of substance left at time t is
$$Q(t) = Q_0 0.96^{t/10}$$
where Q_0 is the initial quantity. We are asked to find the time t at which the quantity of material will be half of the initial quantity. Thus we are asked to solve
$$0.5 Q_0 = Q_0 0.96^{t/10}$$
$$0.5 = 0.96^{t/10}$$
$$\ln 0.5 = \ln 0.96^{t/10}$$
$$(t/10) \ln 0.96 = \ln 0.5$$
$$t = 10 \frac{\ln 0.5}{\ln 0.96} \approx 170$$
Thus the half life of the material is roughly 170 hours.

9. Assuming a rate of inflation of 4.6% a year, prices increase from one year to the next by a factor of 1.046. Letting t be time in years from 1990, the price P of a stamp in dollars is given, in our model, by the equation
$$P(t) = 0.29(1.046)^t,$$

where the 0.29 comes from the condition that $P(t) = 29¢$ at $t = 0$. To find the time when it will cost a dollar to mail a letter, we solve

$$0.29(1.046)^t = 1,$$
$$\log 0.29(1.046)^t = \log 1,$$
$$\log 0.29 + t \log 1.046 = 0,$$
$$t = -\frac{\log 0.29}{\log 1.046} \approx 27.52.$$

So a stamp should cost \$1 by the middle of the year $1990 + 27 = 2017$.

13. $P = 10(e^{0.917})^t = 10(2.5)^t$. Exponential growth because $0.917 > 0$ or $2.5 > 1$.

17. We want $2^t = e^{kt}$ so $2 = e^k$ and $k = \ln 2 = 0.693$. Thus $P = P_0 e^{0.693t}$.

21. (a) We have $P_0 = 1$ million, and $k = 0.02$, so $P(t) = (1{,}000{,}000)(e^{0.02t})$.
    (b)

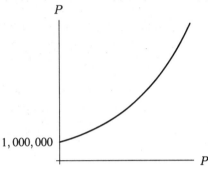

Figure 1.10

25. (a) The pressure P at 6,198 meters is given in terms of the pressure P_0 at sea level to be

$$P = P_0 e^{-1.2 \times 10^{-4} h}$$
$$= P_0 e^{(-1.2 \times 10^{-4})6198}$$
$$= P_0 e^{-0.74376}$$
$$= 0.47532 P_0 \quad \text{or about } 47.5\% \text{ of sea level pressure.}$$

(b) At $h = 12{,}000$ meters, we have

$$P = P_0 e^{-1.2 \times 10^{-4} h}$$
$$= P_0 e^{(-1.2 \times 10^{-4})12{,}000}$$
$$= P_0 e^{-1.44}$$
$$= 0.2369 P_0 \quad \text{or about } 24\% \text{ of sea level pressure.}$$

29. If Q is the amount of strontium-90 which remains at time t, and Q_0 is the original amount, then

$$Q = Q_0 e^{-0.0247t}$$

So after 100 years,

$$Q = Q_0 e^{-0.0247 \cdot 100}$$

and
$$\frac{Q}{Q_0} = e^{-2.47} \approx 0.0846$$
so about 8.46% of the strontium-90 would remain.

Note: If you assume that 2.47% is the *annual* rate, rather than the continuous rate, the answer is not very different:
$$Q = Q_0(1 - 0.0247)^{100} \quad \text{giving} \quad \frac{Q}{Q_0} \approx 0.082 \quad \text{or} \quad 8.2\%.$$

33. (a) $e^{0.06} = 1.0618365$ which means the bank balance has increased by approximately 6.18%.
    (b) $e^{0.06t} = 2$, so $t = \frac{\ln 2}{0.06} = 11.55$ years.
    (c) $e^{rt} = 2$ so $t = \frac{\ln 2}{r}$.

Answers for Section 1.9

1. (a) The equation is $y = 2x^2 + 1$. Note that its graph is narrower than the graph of $y = x^2$ which appears in grey.
    (b) $y = 2(x^2 + 1)$ moves the graph up one unit and *then* stretches it by a factor of two.

    (a)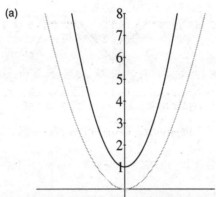
    (b)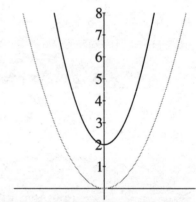

    (c) No, the graphs are not the same. Note that stretching vertically leaves any point whose y-value is zero in the same place but moves any other point. This is the source of the difference because if you stretch it first, its lowest point stays at the origin. Then you shift it up by one and its lowest point is $(0, 1)$. Alternatively, if you shift it first, its lowest point is $(0, 1)$ which, when stretched by 2, becomes $(0, 2)$.

5. (a) $f(g(t)) = f\left(\frac{1}{t+1}\right) = \left(\frac{1}{t+1} + 7\right)^2$
    (b) $g(f(t)) = g((t+7)^2) = \frac{1}{(t+7)^2 + 1}$
    (c) $f(t^2) = (t^2 + 7)^2$
    (d) $g(t-1) = \frac{1}{(t-1)+1} = \frac{1}{t}$

9. $m(z) - m(z-h) = z^2 - (z-h)^2 = 2zh - h^2$.
13. $f(x) = x^3$, $g(x) = \ln x$.
17. Here are the graphs.

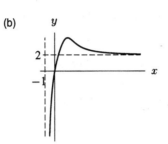

Figure 1.11: $y = 2f(x)$

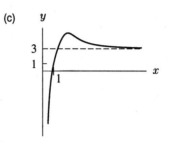

Figure 1.12: $y = f(x+1)$ Figure 1.13: $y = f(x) + 1$

21.

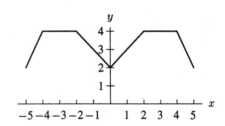

Figure 1.14: Graph of $y = f(x) + 2$

25.

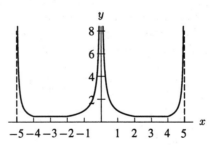

Figure 1.15: Graph of $y = \dfrac{1}{f(x)}$

29. Computing $f(g(x))$ as in Problem 26, we get the following table. From it we graph $f(g(x))$.

TABLE 1.1

x	$g(x)$	$f(g(x))$
-3	0.6	-0.5
-2.5	-1.1	-1.3
-2	-1.9	-1.2
-1.5	-1.9	-1.2
-1	-1.4	-1.3
-0.5	-0.5	-1
0	0.5	-0.6
0.5	1.4	-0.2
1	2	0.4
1.5	2.2	0.5
2	1.6	0
2.5	0.1	-0.7
3	-2.5	0.1

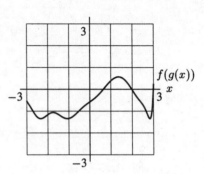

33.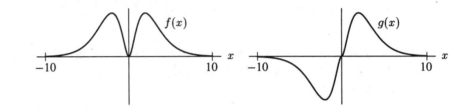

Figure 1.16

Answers for Section 1.10

1. (I) Degree ≥ 3, leading coefficient negative.
   (II) Degree ≥ 4, leading coefficient positive.
   (III) Degree ≥ 4, leading coefficient negative.
   (IV) Degree ≥ 5, leading coefficient negative.
   (V) Degree ≥ 5, leading coefficient positive.

5. We can write $f(x)$ as

$$f(x) = (x-2)(3-x) = -x^2 + 5x - 6.$$

Thus the graph of $f(x)$ will be an upside-down parabola and the function $f(x)$ will equal zero at $x = 2$ and $x = 3$. We also know that

$$f(0) = (-2)(3) = -6,$$

so the graph intersects the y-axis at the point $(0, -6)$. The graph of $f(x)$ looks like

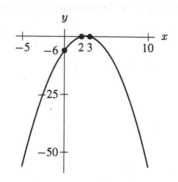

Figure 1.17

9.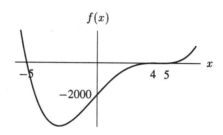

13. $f(x) = x^n$ is even for n an even integer, and odd for n an odd integer.

17. (a) $R(P) = kP(L - P)$, where k is a positive constant.
    (b)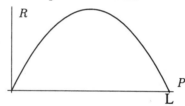

21. (a) $f(x) = k(x + 2)(x - 2)^2(x - 5) = k(x^4 - 7x^3 + 6x^2 + 28x - 40)$, where $k < 0$. ($k \approx -\frac{1}{15}$ if the scales are equal; otherwise one can't tell how large k is.)
    (b) This function is increasing for $x < -1$ and for $2 < x < 4$, decreasing for $-1 < x < 2$ and for $4 < x$.

Answers for Section 1.11

1. From the example, we know that $h = 5 + 4.9\cos(\frac{\pi}{6}t)$, where t represents hours after midnight and h represents the height of the water.

$$\text{At 3:00 am, } t = 3 \text{ so } h = 5 + 4.9\cos\left(\frac{\pi}{6} \cdot 3\right) = 5 + 4.9(0) = 5 \text{ feet}$$

$$\text{At 4:00 am, } t = 4 \text{ so } h = 5 + 4.9\cos\left(\frac{\pi}{6} \cdot 4\right) = 5 + 4.9(-0.5) = 2.55 \text{ feet}$$

$$\text{At 5:00 pm, } t = 17 \text{ so } h = 5 + 4.9\cos\left(\frac{\pi}{6} \cdot 17\right) \approx 5 + 4.9(-0.866) \approx 0.76 \text{ feet}$$

5. The record's period is $\frac{1}{33\frac{1}{3}} = \frac{3}{100}$ minute, or 1.8 seconds.

9.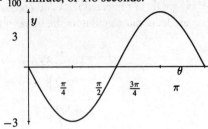

The amplitude is 3; the period is π.

13. This graph is a sine curve with period 8π and amplitude 2, so it is given by $f(x) = 2\sin(\frac{x}{4})$.

17. (a)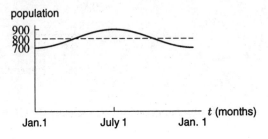

Figure 1.18

(b) Average value of population = $\frac{700+900}{2} = 800$, amplitude = $\frac{900-700}{2} = 100$, and period = 12 months, so $B = 2\pi/12 = \pi/6$. Since the population is at its minimum when $t = 0$, we use a negative cosine:

$$P = 800 - 100\cos\left(\frac{\pi t}{6}\right).$$

21. (a) The period is 2π.

(b) The period of $\sin 3t$ is $\frac{2}{3}\pi$. Period of $\cos t$ is 2π.

(c) The combined function repeats when each part repeats separately—although $\sin 3t$ repeats every $\frac{2}{3}\pi$, the combined function must "wait" until $\cos t$ repeats for it to return to its original value.

Answers for Section 1.12

1. (a)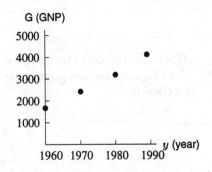

Figure 1.19

A line seems to fit the data somewhat well.

(b) If we let G represent the GNP and y represent the years that have passed since 1960, we get from our calculators the linear regression equation.

$$G = F(y) = 83.65y + 1612$$

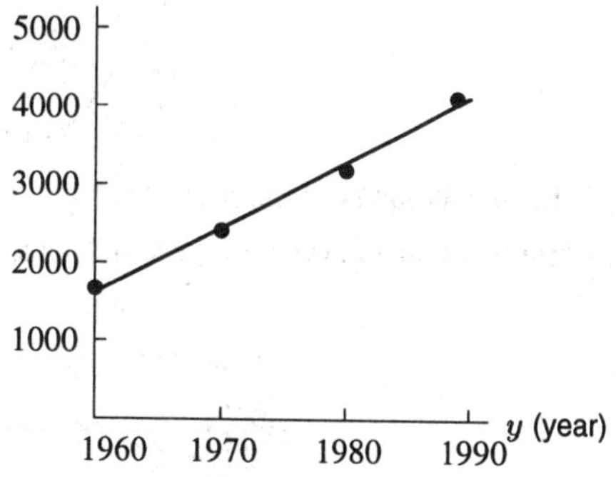

Figure 1.20

(c) Using our equation, we estimate the GNP in 1985 as

$$F(25) = (83.65)(25) + 1612$$
$$\approx 2091 + 1612$$
$$= 3703$$

We can estimate the GNP in 2000 with

$$F(40) = 3346 + 1612$$
$$= 4958$$

We can expect that our estimation will work better for the year 1985 than for the year 2000. This is because this sort of estimation works best the closer the year that we want to estimate is to the existing data points. Thus, our estimation will be more accurate for 1985, which is within the range of the data points given, than 2000, for which we need to extrapolate.

5. (a) The annual growth rate of an exponential function of the form Ar^t is just $(r-1)$, analogous to the rate of interest in interest problems. Thus, in the given function

$$r - 1 = 1.0026 - 1.00$$
$$= 0.0026$$

This means that the CO_2 concentration grows by 0.26% every year.

(b) The CO_2 concentration given by the model for 1900, when we substitute $t = 0$ into the function, is

$$C = 272.27(1.0026)^0$$
$$= 272.27(1)$$
$$= 272.27$$

The CO_2 concentration given by the model for 1980 is, substituting $t = 80$

$$C = 272.27(1.0026)^{80}$$
$$\approx 272.27(1.23)$$
$$\approx 335.1$$

This is not too far from the real concentration of 338.5.

9. (a)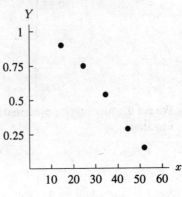

Figure 1.21

An exponential curve seems to provide the best fit. Note that we are looking at a data set with a negative slope.

(b) The best linear regression function for this data set is

$$Y = -0.021x + 1.23$$

The slope of this function -0.021 indicates that the farther the distance from the goal line, the less likely a field goal kick will succeed.

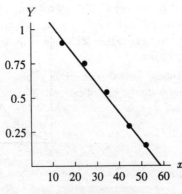

Figure 1.22

(c) The best exponential regression function is

$$Y = 2.17(0.954)^x$$

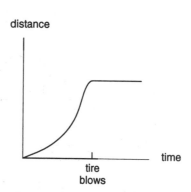

Figure 1.23

We get the success rate predicted by substituting $x = 50$ into the exponential regression function

$$Y = 2.17(0.954)^x$$
$$= 2.17(0.954)^{50}$$
$$\approx 2.17(0.095)$$
$$\approx 0.206$$

Thus, the predicted success rate is 20.6%.
(d) Of the two graphs, the linear model seems to fit the data best.

REVIEW PROBLEMS FOR CHAPTER ONE

1.

```
distance
```

(graph rising slowly then sharply leveling off, with "tire blows" marked on time axis)

time

5. (a) Four zeros, at approximately $x = -4.6, 1.2, 2.7$, and 4.1.
   (b) $f(2)$ is the y-value corresponding to $x = 2$, so $f(2)$ is about -1. Likewise, $f(4)$ is about 0.4.
   (c) Decreasing near $x = -1$, increasing near $x = 3$.
   (d) Concave up near $x = 2$, concave down near $x = -4$.
   (e) Increasing on $x < -1.5$ and on $2 < x < 3.5$.

9. (a) Advertising is generally cheaper in bulk; spending more money will give better and better marginal results initially. (Spending $5,000 could give you a big newspaper ad reaching 200,000 people; spending $100,000 could give you a series of TV spots reaching 50,000,000 people.)

(b) The temperature of a hot object decreases at a rate proportional to the difference between its temperature and the temperature of the air around it. Thus, the temperature of a very hot object decreases more quickly than a cooler object. The graph is decreasing and concave up. (We are assuming that the coffee is all at the same temperature.)

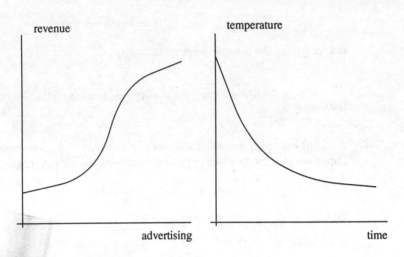

13. One possible graph is given below.

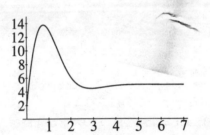

17. Looking at y_1, we see that it is a decreasing function of x. Furthermore, we see that the rate of decrease of y_1 decreases as x increases. Thus, given the choices, y_1 must be inversely proportional to x, so we have

$$y_1 = \frac{k}{x}.$$

We know that at $x = 10$ we have $y_1 = 300$, so

$$300 = \frac{k}{10},$$

and we get for the constant of proportionality

$$k = 3000.$$

Hence,
$$y_1 = \frac{3000}{x}.$$

Looking at y_2, we see that y_2 is an increasing function of x. Also, notice that the rate of increase of y_2 increases as x increases. Therefore, y_2 must be proportional to x^2, so we have

$$y_2 = kx^2.$$

We know that at $x = 10$ we have $y_2 = 200$, so

$$200 = k(10)^2 = 100k,$$

and we get for the constant of proportionality

$$k = 2.$$

Thus,
$$y_2 = 2x^2.$$

Looking at y_3, we see that the rate of increase of y_3 is constant for all given x values. Therefore, y_3 must be linearly proportional to x, and we can write

$$y_3 = kx.$$

We know that at $x = 10$ we have $y_3 = 2.50$, so

$$2.50 = k(10),$$

and we get for the constant of proportionality

$$k = 0.25.$$

Hence,
$$y_3 = 0.25x.$$

21. $P = 2.91(e^{0.55})^t = 2.91(1.733)^t$

25. $e^r = 1.05$ so $r = \ln(1.05) = 0.0488$ or 4.88%.

29. (a) Each day prices are multiplied by 1.001, so after 365 days prices are $(1.001)^{365} \approx 1.44$ what they were originally. Therefore, they increase by about 44% a year.
    (b) Guess about two years, since $1.44^2 \approx 2$. $(1.001)^{2(365)} = (1.001)^{730} \approx 2.074$, so it's a good guess.

33. Since this function has a y-intercept at $(0, 2)$, we expect it to have the form $y = 2e^{kx}$. Again, we find k by forcing the other point to lie on the graph:

$$1 = 2e^{2k}$$
$$\frac{1}{2} = e^{2k}$$
$$\ln\left(\frac{1}{2}\right) = 2k$$
$$k = \frac{\ln(\frac{1}{2})}{2} \approx -0.34657.$$

This value is negative, which makes sense since the graph shows exponential decay. The final equation, then, is
$$y = 2e^{-0.34657x}.$$

37. $x = ky(y-4) = k(y^2 - 4y)$, where $k > 0$ is any constant.

41. (a) is $g(x)$ since it is linear. (b) is $f(x)$ since it has decreasing slope; the slope starts out about 1 and then decreases to about $\frac{1}{10}$. (c) is $h(x)$ since it has increasing slope; the slope starts out about $\frac{1}{10}$ and then increases to about 1.

45. (a) The rate R is the difference of the rate at which the glucose is being injected, which is given to be constant, and the rate at which the glucose is being broken down, which is given to be proportional to the amount of glucose present. Thus we have the formula
$$R = k_1 - k_2 G$$
where k_1 is the rate that the glucose is being injected, k_2 is the constant relating the rate that it is broken down to the amount present, and G is the amount present.

(b)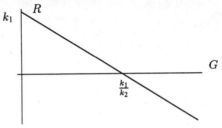

49. Depth $= d = 7 + 1.5 \sin\left(\frac{\pi}{3}t\right)$

CHAPTER 2

Answers for Section 2.1

1. (a) Recall that the velocity is the ratio of the distance travelled and the time spent travelling. Thus, between $t = 0$ and $t = 15$, we get average velocity v:

$$v = \frac{s_{15} - s_0}{15 - 0}$$
$$= \frac{105 - 0}{15}$$
$$= 7 \text{ft/s}$$

(b) Between $t = 10$ and $t = 30$, the car has the average velocity v:

$$v = \frac{s_{30} - s_{10}}{30 - 10}$$
$$= \frac{410 - 55}{20}$$
$$= 17.75 \text{ft/s}$$

(c) Between $t = 10$ and $t = 30$, the distance travelled is

$$d = s_{30} - s_{10}$$
$$= 410 - 55$$
$$= 355 \text{ft}$$

5. (a) The rate of change R is the difference in amounts divided by the change in time.

$$R = \frac{50.5 - 35.6}{1993 - 1987}$$
$$= \frac{14.9}{6}$$
$$= 2.5 \quad \text{billion dollars/yr}$$

This means that in the years between 1987 and 1993, the amount of money spent on tobacco has been increasing at a rate of $2,500,000,000 per year.

(b) To have a negative rate of change, the amount spent has to decrease at one of these 1 year intervals. Looking at the data, one can see that between 1992 and 1993, the amount spent on tobacco products decreased by .4 billion dollars. Thus, the average rate of change is negative between 1992 and 1993.

Answers for Section 2.2

1.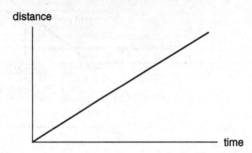

5.
$$\begin{pmatrix} \text{Average velocity} \\ 0 < t < 0.2 \end{pmatrix} = \frac{s(0.2) - s(0)}{0.2 - 0} = \frac{0.5}{0.2} = 2.5 \text{ ft/sec.}$$

$$\begin{pmatrix} \text{Average velocity} \\ 0.2 < t < 0.4 \end{pmatrix} = \frac{s(0.4) - s(0.2)}{0.4 - 0.2} = \frac{1.3}{0.2} = 6.5 \text{ ft/sec.}$$

A reasonable estimate of the velocity at $t = 0.2$ is the average: $\frac{1}{2}(6.5 + 2.5) = 4.5$ ft/sec.

9. The slope is positive at A and D; negative at C and F. The slope is most positive at A; most negative at F.

13. A graph of $f(x)$ is shown in Fig 2.1.

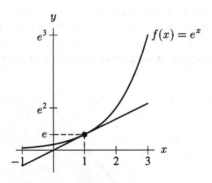

Figure 2.1

Looking at the graph we see that the tangent to $f(x)$ at the point $(1, f(1))$ is a line with positive slope. Since the slope of the tangent is exactly the derivative of $f(x)$ at $x = 1$ we know that the derivative is positive.

17.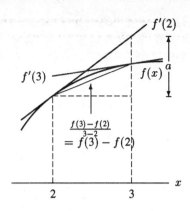

The quantities $f'(2)$, $f'(3)$ and $f(3) - f(2)$ have the following interpretations:
$f'(2)$ = slope of the tangent line at $x = 2$
$f'(3)$ = slope of the tangent line at $x = 3$
$f(3) - f(2) = \frac{f(3)-f(2)}{3-2}$ = slope of the secant line from $f(2)$ to $f(3)$.

From the figure, it is clear that $0 < f(3) - f(2) < f'(2)$. By extending the secant line past the point $(3, f(3))$, we can see that it lies above the tangent line at $x = 3$. Thus $0 < f'(3) < f(3) - f(2) < f'(2)$. From the figure, the height a appears less than 1, so $f'(2) = \frac{a}{3-2} = \frac{a}{1} < 1$.
Thus
$$0 < f'(3) < f(3) - f(2) < f'(2) < 1.$$

21. Notice that we can't get all the information we want just from the graph of f for $0 \leq x \leq 2$, shown above to the left. Looking at this graph, it looks as if the slope at $x = 0$ is 0. But if we zoom in on the graph near $x = 0$, we get the graph of f for $0 \leq x \leq 0.05$, shown above on the upper right. We see that f does dip down quite a bit between $x = 0$ and $x \approx 0.11$. In fact, it now looks like $f'(0)$ is around -1. Note that since $f(x)$ is undefined for $x < 0$, this derivative only makes sense as we approach zero from the right.

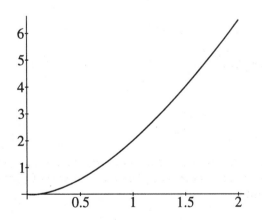
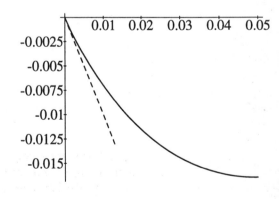

We zoom in on the graph of f near $x = 1$ to get a more accurate picture to estimate $f'(1)$. A graph of f for $0.7 \leq x \leq 1.3$ is shown below. [Keep in mind that the axes shown in this graph don't cross at the origin!] Here we see that $f'(1) \approx 3.5$.

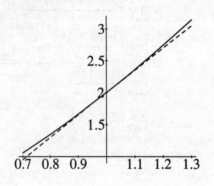

25.
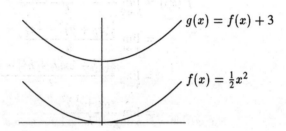

(a) From the figure above, it appears that the slopes of the tangent lines to the two graphs are the same at each x. For $x = 0$, the slopes of the tangents to the graphs of $f(x)$ and $g(x)$ at 0 are

$$f'(0) = \lim_{h \to 0} \frac{f(0+h) - f(0)}{h}$$
$$= \lim_{h \to 0} \frac{f(h) - 0}{h}$$
$$= \lim_{h \to 0} \frac{\frac{1}{2}h^2}{h}$$
$$= \lim_{h \to 0} \frac{1}{2}h$$
$$= 0,$$

$$g'(0) = \lim_{h \to 0} \frac{g(0+h) - g(0)}{h}$$
$$= \lim_{h \to 0} \frac{g(h) - g(0)}{h}$$
$$= \lim_{h \to 0} \frac{\frac{1}{2}h^2 + 3 - 3}{h}$$
$$= \lim_{h \to 0} \frac{\frac{1}{2}h^2}{h}$$
$$= \lim_{h \to 0} \frac{1}{2}h$$
$$= 0.$$

For $x = 2$, the slopes of the tangents to the graphs of $f(x)$ and $g(x)$ are

$$f'(2) = \lim_{h \to 0} \frac{f(2+h) - f(2)}{h}$$
$$= \lim_{h \to 0} \frac{\frac{1}{2}(2+h)^2 - \frac{1}{2}(2)^2}{h}$$
$$= \lim_{h \to 0} \frac{\frac{1}{2}(4 + 4h + h^2) - 2}{h}$$
$$= \lim_{h \to 0} \frac{2 + 2h + \frac{1}{2}h^2 - 2}{h}$$
$$= \lim_{h \to 0} \frac{2h + \frac{1}{2}h^2}{h}$$
$$= \lim_{h \to 0} \left(2 + \frac{1}{2}h\right)$$
$$= 2,$$

$$g'(2) = \lim_{h \to 0} \frac{g(2+h) - g(2)}{h}$$
$$= \lim_{h \to 0} \frac{\frac{1}{2}(2+h)^2 + 3 - (\frac{1}{2}(2)^2 + 3)}{h}$$
$$= \lim_{h \to 0} \frac{\frac{1}{2}(2+h)^2 - \frac{1}{2}(2)^2}{h}$$
$$= \lim_{h \to 0} \frac{\frac{1}{2}(4 + 4h + h^2) - 2}{h}$$
$$= \lim_{h \to 0} \frac{2 + 2h + \frac{1}{2}(h^2) - 2}{h}$$
$$= \lim_{h \to 0} \frac{2h + \frac{1}{2}(h^2)}{h}$$
$$= \lim_{h \to 0} \left(2 + \frac{1}{2}h\right)$$
$$= 2.$$

For $x = x_0$, the slopes of the tangents to the graphs of $f(x)$ and $g(x)$ are

$$f'(x_0) = \lim_{h \to 0} \frac{f(x_0 + h) - f(x_0)}{h}$$
$$= \lim_{h \to 0} \frac{\frac{1}{2}(x_0 + h)^2 - \frac{1}{2}x_0^2}{h}$$
$$= \lim_{h \to 0} \frac{\frac{1}{2}(x_0^2 + 2x_0 h + h^2) - \frac{1}{2}x_0^2}{h}$$
$$= \lim_{h \to 0} \frac{x_0 h + \frac{1}{2}h^2}{h}$$
$$= \lim_{h \to 0} \left(x_0 + \frac{1}{2}h\right)$$
$$= x_0,$$

$$g'(x_0) = \lim_{h \to 0} \frac{g(x_0 + h) - g(x_0)}{h}$$
$$= \lim_{h \to 0} \frac{\frac{1}{2}(x_0 + h)^2 + 3 - (\frac{1}{2}(x_0)^2 + 3)}{h}$$
$$= \lim_{h \to 0} \frac{\frac{1}{2}(x_0 + h)^2 - \frac{1}{2}(x_0)^2}{h}$$
$$= \lim_{h \to 0} \frac{\frac{1}{2}(x_0^2 + 2x_0 h + h^2) - \frac{1}{2}x_0^2}{h}$$
$$= \lim_{h \to 0} \frac{x_0 h + \frac{1}{2}h^2}{h}$$
$$= \lim_{h \to 0} \left(x_0 + \frac{1}{2}h\right)$$
$$= x_0.$$

(b)

$$g'(x) = \lim_{h \to 0} \frac{g(x+h) - g(x)}{h}$$
$$= \lim_{h \to 0} \frac{f(x+h) + C - (f(x) + C)}{h}$$
$$= \lim_{h \to 0} \frac{f(x+h) - f(x)}{h}$$
$$= f'(x).$$

Answers for Section 2.3

1.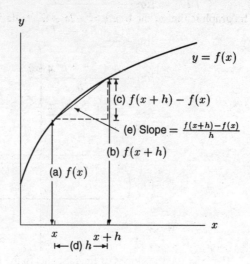

5. Substituting smaller and smaller values of h gives Table 2.1.

    TABLE 2.1

h	$\frac{\ln(2+h)-\ln(2)}{h}$
0.1	0.4879
0.01	0.4988
0.001	0.4999

    From the table we can conclude that the limit, correct to one decimal place, is

    $$\lim_{h \to 0} \frac{\ln(2+h) - \ln(2)}{h} = 0.5$$

9.
$$g'(-1) = \lim_{h \to 0} \frac{g(-1+h) - g(-1)}{h} = \lim_{h \to 0} \frac{(3(-1+h)^2 + 5(-1+h)) - (3(-1)^2 + 5(-1))}{h}$$
$$= \lim_{h \to 0} \frac{(3(1 - 2h + h^2) - 5 + 5h) - (-2)}{h} = \lim_{h \to 0} \frac{3 - 6h + 3h^2 - 3 + 5h}{h}$$
$$= \lim_{h \to 0} \frac{(-h + 3h^2)}{h} = \lim_{h \to 0}(-1 + 3h) = -1.$$

13.

TABLE 2.2

x	2.998	2.999	3.000	3.001	3.002
$x^3 + 4x$	38.938	38.969	39.000	39.031	39.062

We see that each x increase of 0.001 leads to an increase in $f(x)$ by about 0.031, so $f'(3) \approx \frac{0.031}{0.001} = 31$.

Answers for Section 2.4

1. The graph is that of the line $y = -2x + 2$. Its derivative is -2.

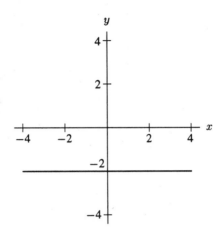

5.

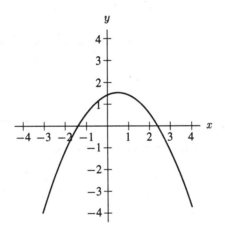

9. We know that $f'(x) \approx \dfrac{f(x+h) - f(x)}{h}$. For this problem, we'll take the average of the values obtained for $h = 1$ and $h = -1$; that's the average of $f(x+1) - f(x)$ and $f(x) - f(x-1)$ which equals $\dfrac{f(x+1) - f(x-1)}{2}$. Thus, $f'(0) \approx f(1) - f(0) = 13 - 18 = -5$.
$f'(1) \approx [f(2) - f(0)]/2 = [10 - 18]/2 = -4$.
$f'(2) \approx [f(3) - f(1)]/2 = [9 - 13]/2 = -2$.
$f'(3) \approx [f(4) - f(2)]/2 = [9 - 10]/2 = -0.5$.
$f'(4) \approx [f(5) - f(3)]/2 = [11 - 9]/2 = 1$.
$f'(5) \approx [f(6) - f(4)]/2 = [15 - 9]/2 = 3$.
$f'(6) \approx [f(7) - f(5)]/2 = [21 - 11]/2 = 5$.
$f'(7) \approx [f(8) - f(6)]/2 = [30 - 15]/2 = 7.5$.
$f'(8) \approx f(8) - f(7) = 30 - 21 = 9$.
The rate of change of $f(x)$ is positive for $4 \le x \le 8$, negative for $0 \le x \le 3$. The rate of change is greatest at about $x = 8$.

13.
$$g'(x) = \lim_{h \to 0} \frac{g(x+h) - g(x)}{h} = \lim_{h \to 0} \frac{2(x+h)^2 - 3 - (2x^2 - 3)}{h}$$
$$= \lim_{h \to 0} \frac{2(x^2 + 2xh + h^2) - 3 - 2x^2 + 3}{h} = \lim_{h \to 0} \frac{4xh + 2h^2}{h}$$
$$= \lim_{h \to 0} (4x + 2h) = 4x$$

17. Since $f'(x) > 0$ for $x < -1$, $f(x)$ is increasing on this interval.
Since $f'(x) < 0$ for $x > -1$, $f(x)$ is decreasing on this interval.
Since $f'(x) = 0$ at $x = -1$, the tangent to $f(x)$ is horizontal at $x = -1$.
One of many possible shapes of $y = f(x)$ is shown in Figure 2.2.

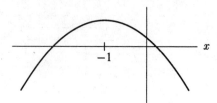

Figure 2.2

21.

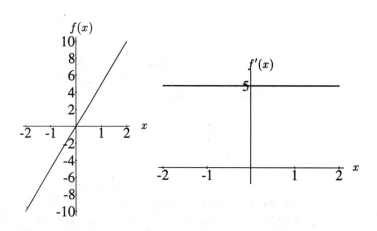

25.

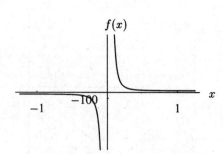

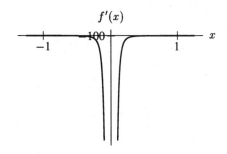

29.

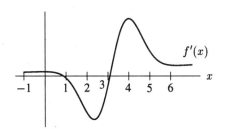

33. If $f(x)$ is even, its graph is symmetric about the y-axis. So the tangent line to f at $x = x_0$ is the same as that at $x = -x_0$ reflected about the y-axis.

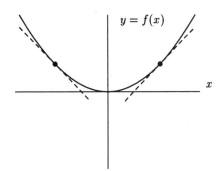

So the slopes of these two tangent lines are opposite in sign, so $f'(x_0) = -f'(-x_0)$, and f' is odd.

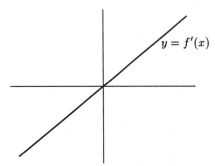

Answers for Section 2.5

1. (a) Velocity is zero at points A, C, F, and H.
   (b) These are points where the acceleration is zero, and hence where the particle switches from speeding up to slowing down or vice versa.

5. Units of $C'(r)$ are dollars/percent. Approximately, $C'(r)$ means the additional amount needed to pay off the loan when the interest rate is increased by 1%. The sign of $C'(r)$ is positive, because increasing the interest rate will increase the amount it costs to pay off a loan.

9. (Note that we are considering the average temperature of the yam, since its temperature is different at different points inside it.)
   (a) It is positive, because the temperature of the yam increases the longer it sits in the oven.
   (b) The units of $f'(20)$ are °F/min. $f'(20) = 2$ means that at time $t = 20$ minutes, the temperature T increases by approximately 2°F for each additional minute in the oven.

13. (a)

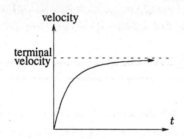

   (b) The graphs should be concave down because wind resistance decreases your acceleration as you speed up, and so the slope of the graph of velocity is decreasing.
   (c) The slope represents the acceleration due to gravity.

17. (a) Clearly the population of the US at any instant is an integer that varies up and down every few seconds as a child is born, a person dies, or a new immigrant arrives. Since these events cannot usually be assigned to an exact instant, the population of the US at any given moment might actually be indeterminate. If we count in units of a thousand, however, the population appears to be a smooth function that has been rounded to the nearest thousand.
   Major land acquisitions such as the Louisiana Purchase caused larger jumps in the population, but since the census is taken only every ten years and the territories acquired were rather sparsely populated, we cannot see these jumps in the census data.
   (b) We can regard rate of change of the population for a particular time t as representing an estimate of how much the population will increase during the year after time t.
   (c) Many economic indicators are treated as smooth, such as the Gross National Product, the Dow Jones Industrial Average, volumes of trading, and the price of commodities like gold. But these figures only change in increments, and not continuously.

Answers for Section 2.6

1. (a) increasing, concave up.
   (b) decreasing, concave down

5. (a)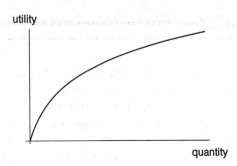

(b) As a function of quantity, utility is increasing but at a decreasing rate; the graph is increasing but concave down. So the derivative of utility is positive, but the second derivative of utility is negative.

9. (a) The EPA will say that the rate of discharge is still rising. The industry will say that the rate of discharge is increasing less quickly, and may soon level off or even start to fall.

(b) The EPA will say that the rate at which pollutants are being discharged is levelling off, but not to zero — so pollutants will continue to be dumped in the lake. The industry will say that the rate of discharge has decreased significantly.

Answers for Section 2.7

1. At $q = 50$, the slope of the revenue is larger than the slope of the cost. Thus, at $q = 50$, marginal revenue is greater than marginal cost and the 50th bus should be made. At $q = 100$ the slope of revenue is less than the slope of cost. Thus, at $q = 100$ the marginal revenue is less than marginal cost and the 100th bus should not be made.

5.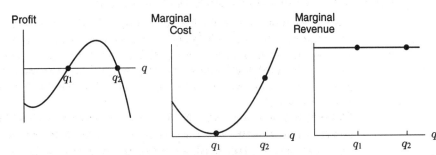

9. The profit is maximized at the point where the difference between revenue and cost is greatest. Thus the profit is maximized at approximately $q = 4000$.

13. (a) We can approximate $C(16)$ by adding $C'(15)$ to $C(15)$. ($C'(15)$ is an estimate of the cost of the 16th item.)

$$C(15) + C'(15) = \$2300 + \$108 = \$2408.$$

(b) Likewise, we can approximate $C(14)$ by subtracting $C'(15)$ from $C(15)$, (where $C'(15)$ is an approximation of the cost of producing the 15th item.)

$$C(15) - C'(15) = \$2300 - \$108 = \$2192.$$

17. Drawing in the tangent line to the point $(600, R(600))$ we get Figure 2.3.

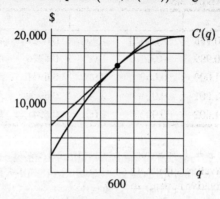

Figure 2.3

We see that for each vertical increase of 2500 in the tangent line gives a corresponding horizontal increase of roughly 150. Thus the marginal revenue at the production level of 600 units is

$$R'(600) = \frac{\text{slope of tangent line}}{\text{to } R(q) \text{ at } q = 600} = \frac{2500}{150} = \frac{5}{3}.$$

This tells us that after the producing 600 units, the revenue for producing the 601st product will be roughly $1.67.

REVIEW PROBLEMS FOR CHAPTER TWO

1.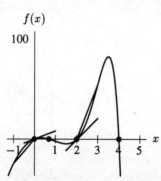

   (a) Using our graphing calculator, we can see four different zeroes, at $x = 0$, $x = 2$, $x = 4$ and $x \approx .7$. This gives us four zeroes in the interval.
   (b) At $x = 0$ and $x = 2$, we see that the tangent has positive slope so f is increasing.
   At $x = 4$, we notice that the tangent to the curve has negative slope, so f is decreasing.
   (c) Comparing the slope of the secant lines at these values, we can see that the average rate of change of f is greater during the interval $2 \leq x \leq 3$.
   (d) Looking at the tangents of the function at $x = 0$ and $x = 2.$, we can see that the slope of the tangent at $x = 2$ is greater. Thus, the instantaneous rate of change of f is greater at $x = 2$.

5.

TABLE 2.3

x	$\ln x$	x	$\ln x$	x	$\ln x$	x	$\ln x$
0.998	−0.0020	1.998	0.6921	4.998	1.6090	9.998	2.3024
0.999	−0.0010	1.999	0.6926	4.999	1.6092	9.999	2.3025
1.000	0.0000	2.000	0.6931	5.000	1.6094	10.000	2.3026
1.001	0.0010	2.001	0.6936	5.001	1.6096	10.001	2.3027
1.002	0.0020	2.002	0.6941	5.002	1.6098	10.002	2.3028

At $x = 1$, the values of $\ln x$ are increasing by 0.001 for each increase in x of 0.001, so the derivative appears to be 1. At $x = 2$, the increase is 0.0005 for each increase of 0.001, so the derivative appears to be 0.5. At $x = 5$, $\ln x$ increases by 0.0002 for each increase of 0.001 in x, so the derivative appears to be 0.2. And at $x = 10$, the increase is 0.0001 over intervals of 0.001, so the derivative appears to be 0.1. These values suggest an inverse relationship between x and $f'(x)$, namely $f'(x) = \frac{1}{x}$.

9.

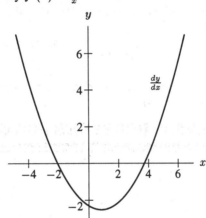

13.

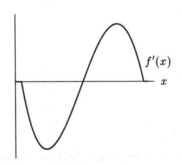

17. (a) $f'(6.75) \approx \frac{f(7.0)-f(6.5)}{7.0-6.5} = \frac{8.2-10.3}{0.5} = -4.2$.

$f'(7.0) \approx \frac{f(7.5)-f(6.5)}{7.5-6.5} = \frac{6.5-10.3}{1.0} = -3.8$.

$f'(8.5) \approx \frac{f(9.0)-f(8.0)}{9.0-8.0} = \frac{3.2-5.2}{1.0} = -2.0$.

(b) To estimate f'' at 7, we should have values for f' at points near 7. We know from (a) that $f'(6.75) \approx -4.2$. Next, estimate $f'(7.25) \approx \frac{6.5-8.2}{0.5} = -3.4$. Then $f''(7) \approx \frac{f'(7.25)-f'(6.75)}{0.5} \approx \frac{-3.4-(-4.2)}{0.5} = 1.6$.

(c) $y - 8.2 = -3.8(x - 7)$ or $y = -3.8x + 34.8$.

(d) We may use the tangent line from (c) to approximate $f(6.8)$. In this case we get

$$y \approx -3.8x + 34.8 = 8.96.$$

[We may also estimate $f(6.8)$ by assuming that the graph of f is straight between the given points $(6.5, 10.3)$ and $(7.0, 8.2)$. This line has the equation $y = -4.2(x - 6.5) + 10.3$ and passes through $(6.8, 9.04)$, so we may estimate $f(6.8) \approx 9.04$. Here, we approximate using the secant line rather than the tangent line.]

As we can see, the two estimates are fairly close.

21. (a)

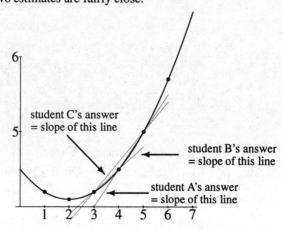

(b) The slope of f appears to be somewhere between student A's answer and student B's, so student C's answer, halfway in between, is probably the most accurate.

(c) Student A's estimate is $f'(x) \approx \frac{f(x+h)-f(x)}{h}$.

Student B's estimate is $f'(x) \approx \frac{f(x)-f(x-h)}{h}$.

Student C's estimate is the average of these two, or

$$f'(x) \approx \frac{1}{2}\left[\frac{f(x+h)-f(x)}{h} + \frac{f(x)-f(x-h)}{h}\right] = \frac{f(x+h)-f(x-h)}{2h}.$$

This estimate is the slope of the chord connecting $(x-h, f(x-h))$ to $(x+h, f(x+h))$.

Thus we estimate that the tangent to a curve is nearly parallel to a chord connecting points h units to the right and left.

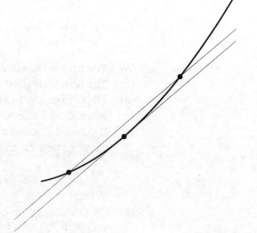

25. (a) See (b).

(b)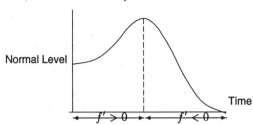

(c) f' is the rate at which the concentration is increasing or decreasing. f' is positive at the start of the disease and negative toward the end. In practice, of course, one cannot measure f' directly. Checking the value of C in blood samples taken on consecutive days would tell us
$$\frac{f(t+1) - f(t)}{(t+1) - t},$$
which is our estimate of $f'(t)$.

29. (a) The population varies periodically with a period of 12 months (i.e. one year).

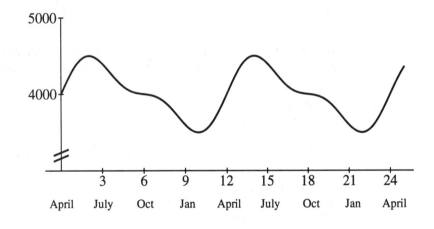

(b) The herd is largest about June 1st when there are about 4500 deer.
(c) The herd is smallest about February 1st when there are about 3500 deer.
(d) The herd grows the fastest about April 1st. The herd shrinks the fastest about July 20 and again about November 15.
(e) It grows the fastest about April 1st when the rate of growth is about 400 deer/month, i.e about 13 new fawns per day.

CHAPTER 3

Answers for Section 3.1

1. (a) Lower estimate $= 60 + 40 + 25 + 10 + 0 = 135$ feet. Upper estimate $= 88 + 60 + 40 + 25 + 10 = 223$ feet.

    (b)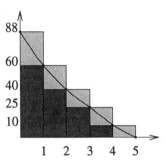

5. Using whole grid squares, we can overestimate the area as $3+3+3+3+2+1 = 15$, and we can underestimate the area as $1+2+2+1+0+0 = 6$. Using triangles as in Figure 3.1, we can overestimate the area as $2 + 2\frac{7}{8} + 3 + 2\frac{1}{2} + 1\frac{1}{2} + \frac{3}{4} = 12\frac{5}{8}$ and we can underestimate the area as $1\frac{1}{2} + 2\frac{1}{4} + 2\frac{1}{2} + 2 + 1 + \frac{1}{4} = 9\frac{1}{2}$. It also appears from the graph that our upper estimate is closer than the lower estimate to the actual area, so we can further estimate the area to be a little greater than $\frac{1}{2}(9\frac{1}{2} + 12\frac{5}{8}) = 11\frac{1}{16}$.

    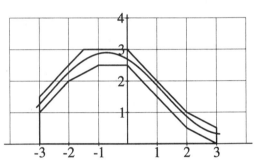

    Figure 3.1: Estimating the Area

9. (a)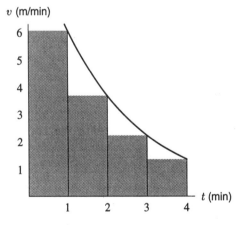

    Figure 3.2

$d = 6 + 3.7 + 2.2 + 1.3 = 13.2$, this is an underestimate.

(b) You travel farther in the first second than between $t = 2$ and $t = 3$, because you are traveling faster in the first second and can cover more ground.

Answers for Section 3.2

1.

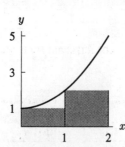

Figure 3.3: Left-hand sum with $n = 2$

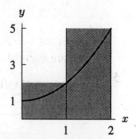

Figure 3.4: Right-hand sum with $n = 2$

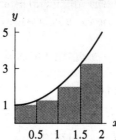

Figure 3.5: Left-hand sum with $n = 4$

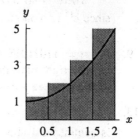

Figure 3.6: Right-hand sum with $n = 4$

For $n = 2$, the left-hand sum is equal to

$$f(0) \cdot 1 + f(1) \cdot 1 = 1 \cdot 1 + 2 \cdot 1$$
$$= 1 + 2$$
$$= 3$$

For $n = 2$, the right-hand sum is equal to

$$f(1) \cdot 1 + f(2) \cdot 1 = 2 \cdot 1 + 5 \cdot 1$$
$$= 2 + 5$$
$$= 7$$

For $n = 4$, the left-hand sum is equal to

$$f(0) \cdot \frac{1}{2} + f\left(\frac{1}{2}\right) \cdot \frac{1}{2} + f(1)\frac{1}{2} + f\left(1\frac{1}{2}\right)\frac{1}{2} = 1 \cdot \frac{1}{2} + \frac{5}{4} \cdot \frac{1}{2} + 2 \cdot \frac{1}{2} + \frac{13}{4} \cdot \frac{1}{2}$$
$$= \frac{1}{2} + \frac{5}{8} + 1 + \frac{13}{8}$$
$$= \frac{15}{4}$$
$$= 3.75$$

For $n = 4$, the right-hand sum is equal to

$$f\left(\frac{1}{2}\right) \cdot \frac{1}{2} + f(1) \cdot \frac{1}{2} + f\left(1\frac{1}{2}\right)\frac{1}{2} + f(2)\frac{1}{2} = \frac{5}{4} \cdot \frac{1}{2} + 2 \cdot \frac{1}{2} + \frac{13}{4} \cdot \frac{1}{2} + 5 \cdot \frac{1}{2}$$
$$= \frac{5}{8} + 1 + \frac{13}{8} + \frac{5}{2}$$
$$= \frac{23}{4}$$
$$= 5.75$$

In both the $n = 2$ and $n = 4$ cases, the left-hand sums are underestimates and the right-hand sums are overestimates.

5. **TABLE 3.1**

n	2	10	50	250
Left-hand Sum	1.14201	1.38126	1.44565	1.45922
Right-hand Sum	2.00115	1.55309	1.48002	1.46610

There is no obvious guess as to what the limiting sum is. We can only observe that since e^{t^2} is monotonic on $[0, 1]$, the true value is between 1.45922 and 1.46610.

9. For $n = 110$, LHS ≈ 4.810 (rounding down) and RHS ≈ 4.905 (rounding up). Since the left and right sums differ by 0.095, their average must be within 0.0475 of the true value, so $\int_1^5 (\ln x)^2\, dx = 4.858$ to one decimal place.

13. Since e^{-t^2} is an even function, $\int_{-3}^3 e^{-t^2}\, dt = 2\int_0^3 e^{-t^2}\, dt$. This is because the integrand is symmetrical about the y-axis, and by symmetry, $\int_{-3}^0 e^{-t^2}\, dt = \int_0^3 e^{-t^2}\, dt$. Since e^{-t^2} is decreasing monotonically on the interval $0 < t < 3$, we can use our error estimation techniques to approximate $\int_0^3 e^{-t^2}\, dt$. Since we want our error to be less than 0.05 for the interval $-3 < t < 3$, we want our error to be less than $\frac{1}{2}(0.05) = 0.025$ for the interval $0 < t < 3$. For $n = 70$, LHS ≈ 0.908 (rounding up) and RHS ≈ 0.864 (rounding down). Since the left and right sums differ by 0.044, their average must be within 0.022 of the true value, so $\int_{-3}^3 e^{-t^2}\, dt = 2\int_0^3 e^{-t^2}\, dt = 2(0.886) = 1.772$, to one decimal place.

17. As in Figure 3.7, the left- and right-hand sums are both equal to $(4\pi) \cdot 3 = 12\pi$, while the integral is smaller. Thus we have:

$$\int_0^{4\pi} (2 + \cos x)\, dx < \text{left-hand sum} = \text{right-hand sum}.$$

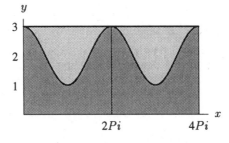

Figure 3.7: Integral vs. Left- and Right-Hand Sums

Answers for Section 3.3

1. The integral represents the area below the graph of $f(x)$ but above the x-axis. Since each square has area 1, by counting squares and half-squares we find
$$\int_1^6 f(x)\,dx = 8.5.$$

5. Calculating both the LHS and RHS and averaging the two, we get
$$\frac{1}{2}(5(100 + 82 + 69 + 60 + 53) + 5(82 + 69 + 60 + 53 + 49)) = 1692.5$$

9.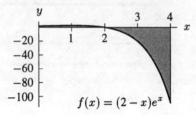

Figure 3.8

More area appears to be below the x-axis than above, so the integral is negative.

13.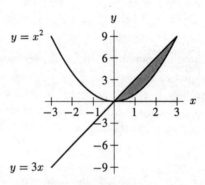

Figure 3.9

Inspection of the graph tells us that the functions intersect at $(0,0)$ and $(3,9)$, with $3x > x^2$ on this interval, so we can find the area by evaluating the integral
$$\int_0^3 (3x - x^2)\,dx.$$

Estimation using Riemann sums gives
$$\int_0^3 (3x - x^2)\,dx = 4.5.$$

So the area between the graphs is 4.5.

17. $\int_{-1}^{1} e^{x^2}\,dx > 0$, since $e^{x^2} > 0$, and $\int_{-1}^{1} e^{x^2}\,dx$ represents the area below the curve $y = e^{x^2}$.

21. (a) We know that $\int_2^5 f(x)\,dx = \int_0^5 f(x)\,dx - \int_0^2 f(x)\,dx$. By symmetry, $\int_0^2 f(x)\,dx = \frac{1}{2}\int_{-2}^2 f(x)\,dx$, so $\int_2^5 f(x)\,dx = \int_0^5 f(x)\,dx - \frac{1}{2}\int_{-2}^2 f(x)\,dx$.
    (b) $\int_2^5 f(x)\,dx = \int_{-2}^5 f(x)\,dx - \int_{-2}^2 f(x)\,dx = \int_{-2}^5 f(x)\,dx - 2\int_{-2}^0 f(x)\,dx$.
    (c) Using symmetry again, $\int_0^2 f(x)\,dx = \frac{1}{2}(\int_{-2}^5 f(x)\,dx - \int_2^5 f(x)\,dx)$.

Answers for Section 3.4

1. $\int_1^6 f(x)\,dx = 8.5$
   average value of $f = \frac{8.5}{5} = 1.7$

5. By a visual estimate, the average value of the left graph is ≈ 8 and the average value of the right graph is ≈ -3.

9. (a) At the end of one hour $t = 60$, and $H = 22°C$.
   (b)
   $$\text{Average temperature} = \frac{1}{60}\int_0^{60}(20 + 980e^{-0.1t})\,dt$$
   $$= \frac{1}{60}(10976) = 183°C.$$

   (c) Average temperature at beginning and end of hour $= (1000 + 22)/2 = 511°C$. The average found in part (b) is smaller than the average of these two temperatures because the bar cools quickly at first and so spends less time at high temperatures. Alternatively, the graph of H against t is concave up.

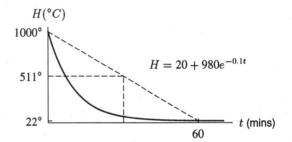

13. (a) $Q(10) = 4e^{-0.036(40)} = 2.8$
    $Q(20) = 4e^{-0.036(20)} = 1.9$
    (b)
    $$\frac{Q(10) + Q(20)}{2} = \frac{2.8 + 1.9}{2} = 2.35$$
    (c) Average value of Q over the interval is 2.30.
    (d) Because Q is concave down, the area under the curve is less than what is obtained by averaging the endpoints.

Answers for Section 3.5

1. $\int_a^b f(t)\,dt$ is measured in
   $$\left(\frac{\text{miles}}{\text{hours}}\right) \cdot (\text{hours}) = \text{miles}.$$

5. For any t, consider the interval $[t, t + dt]$. During this interval, oil is leaking out at an approximately constant rate of $f(t)$ gallons/minute. Thus, the amount of oil which has leaked out during this interval can be expressed as

$$\text{Amount of oil leaked} = \text{Rate} \times \text{Time} = f(t)\, dt$$

and the units of $f(t)\, dt$ are gallons/minute × minutes = gallons. The total amount of oil leaked is obtained by adding all these amounts between $t = 0$ and $t = 60$. (An hour is 60 minutes.) The sum of all these infinitesimal amounts is the integral

$$\begin{array}{c}\text{Total amount of} \\ \text{oil leaked, in gallons}\end{array} = \int_0^{60} f(t)\, dt.$$

9. (a) The amount leaked between $t = 0$ and $t = 2$ is $\int_0^2 R(t)\, dt$.
   (b)
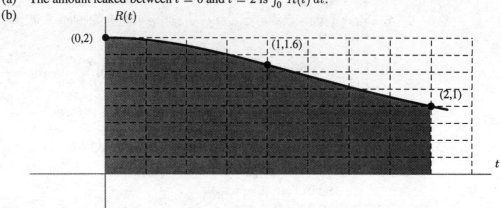

(c) The rectangular boxes on the diagram each have area $\frac{1}{16}$. Of these 45 are wholly beneath the curve, hence the area under the curve is certainly more than $\frac{45}{16} = 2\frac{13}{16} > 2.81$. There are 10 more partially beneath the curve, and so the desired area is completely covered by 55 boxes. Therefore the area is less than $\frac{55}{16} = 3\frac{7}{16} < 3.44$.

These are very safe estimates but far apart. We can do much better by estimating what fractions of the broken boxes are beneath the curve. Using this method, we can estimate the area to be about 3.2, which corresponds to 3.2 gallons leaking over two hours.

13. The total cost in producing 20 units is given by

$$500 + \int_0^{20} \frac{10 e^{0.08q}}{q + 1}\, dq.$$

The 500 is the fixed cost associated with production and the integral is the variable cost.

$$500 + \int_0^{20} \frac{10 e^{0.08q}}{q + 1}\, dq = 500 + 52.8$$

so the total cost is approximately $552.80.

17. Notice that the area of a square on the graph represents $\frac{10}{6}$ miles. At $t = 1/3$ hours, $v = 0$. The area between the curve v and the t-axis over the interval $0 \le t \le 1/3$ is $-\int_0^{1/3} v\, dt \approx \frac{5}{3}$.

v is negative here, so she is moving toward the lake. At $t = \frac{1}{3}$, she is about $5 - \frac{5}{3} = \frac{10}{3}$ miles from the lake. Then, as she moves away from the lake, v is positive for $\frac{1}{3} \le t \le 1$. At $t = 1$,

$$\int_0^1 v\,dt = \int_0^{1/3} v\,dt + \int_{1/3}^1 v\,dt \approx -\frac{5}{3} + 8 \cdot \frac{10}{6} = \frac{35}{3},$$

and the cyclist is about $5 + \frac{35}{3} = \frac{50}{3} = 16\frac{2}{3}$ miles from the lake. Since, starting from the moment $t = \frac{1}{3}$, she moves away from the lake, the cyclist will be farthest from the lake at $t = 1$. The maximal distance equals $16\frac{2}{3}$ miles.

Answers for Section 3.6

1. Since $F(0) = 0$, $F(b) = \int_0^b f(t)\,dt$. For each b we determine $F(b)$ graphically as follows:
   $F(0) = 0$
   $F(1) = F(0) + \text{Area of } 1 \times 1 \text{ rectangle} = 0 + 1 = 1$
   $F(2) = F(1) + \text{Area of triangle } (\frac{1}{2} \cdot 1 \cdot 1) = 1 + 0.5 = 1.5$
   $F(3) = F(2) + \text{Negative of area of triangle} = 1.5 - 0.5 = 1$
   $F(4) = F(3) + \text{Negative of area of rectangle} = 1 - 1 = 0$
   $F(5) = F(4) + \text{Negative of area of rectangle} = 0 - 1 = -1$
   $F(6) = F(5) + \text{Negative of area of triangle} = -1 - 0.5 = -1.5$
   The graph of $F(t)$, for $0 \le t \le 6$, is shown in Figure 3.10.

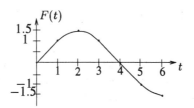

Figure 3.10

5. Since $F'(x)$ is positive for $0 < x < 2$ and $F'(x)$ is negative for $2 < x < 2.5$, $F(x)$ increases on $0 < x < 2$ and decreases on $2 < x < 2.5$. Form this we conclude that $F(x)$ has a maximum at $x = 2$. From the process used in Problem 3 we see that the chart agrees with this assumption and that $F(2) = 5.333$.

9. First rewrite each of the quantities in terms of f', since we have the graph of f'. If A_1 and A_2 are the positive areas shown in Figure 3.11:

$$f(3) - f(2) = \int_2^3 f'(t)\,dt = -A_1$$

$$f(4) - f(3) = \int_3^4 f'(t)\,dt = -A_2$$

$$\frac{f(4) - f(2)}{2} = \frac{1}{2}\int_2^4 f'(t)\,dt = -\frac{A_1 + A_2}{2}$$

Since Area $A_1 >$ Area A_2,

$$A_2 < \frac{A_1 + A_2}{2} < A_1$$

so

$$-A_1 < -\frac{A_1 + A_2}{2} < -A_2$$

and therefore

$$f(3) - f(2) < \frac{f(4) - f(2)}{2} < f(4) - f(3).$$

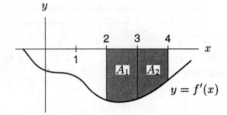

Figure 3.11

13.

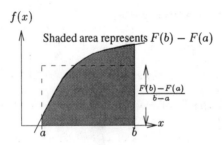

Note that we are using the interpretation of the definite integral as the length of the interval times the average value of the function on that interval, which we developed in Section 3.3.

REVIEW PROBLEMS FOR CHAPTER THREE

1. For $n = 210$, the left sum ≈ 1.466 (rounding up) and the right sum ≈ 1.417 (rounding down). Since the left and right sums differ by 0.049, their average must be within 0.245 of the true value, so the value of the integral will be 1.442 to one decimal place accuracy.

5. For $n = 10$, LEFT ≈ -0.086 (rounding down) and RIGHT ≈ -0.080 (rounding up). Since the left and right sums differ by 0.006, their average must be within 0.003 of the true value, so $\int_2^3 -\frac{1}{(r+1)^2}\,dx = -0.083$ to one decimal place.

9. (a) Since $f(t) = 2t + 3$ is strictly increasing on $1 \leq t \leq 4$, we can use the error formula

$$0.1 > |f(4) - f(1)| \cdot \Delta t$$
$$0.1 > |11 - 5| \cdot \Delta t$$
$$\Delta z < \frac{0.1}{6} = 0.0167$$

So we should use $n = \frac{4-1}{0.0167} = 180$ intervals. With $n = 180$, $LHS \approx 23.95$ and $RHS \approx 24.05$. Since these sums differ by 0.1. their average must be accurate within 0.05 of the true value, so

$$\int_1^4 (2t + 3)\,dt = \frac{23.95 + 24.05}{2} = 24$$

within one decimal place.

(b) Using the Fundamental Theorem of Calculus with $a = 1$ and $b = 4$ we get

$$\int_1^4 (2t + 3)\,dt = F(4) - F(1) = (4^2 + 12) - (1^2 + 3) = 28 - 4 = 24.$$

This is the exact value of the integral.

13. (a)
$$\text{Average population} = \frac{1}{10}\int_0^{10} 67.38(1.026)^t\,dt$$

Evaluating the integral numerically gives

$$\text{Average population} \approx 76.8 \text{ million}$$

(b) In 1980, $t = 0$, and $P = 67.38(1.026)^0 = 67.38$.
In 1990, $t = 10$, and $P = 67.38(1.026)^{10} = 87.10$.
Average $= \frac{1}{2}(67.38 + 87.10) = 77.24$ million.

(c) If P had been linear, the average value found in (a) would have been the one we found in (b). Since the population graph is concave up, it is below the secant line. Thus, the actual values of P are less than the corresponding values on the secant line, and so the average found in (a) is smaller than that in (b).

17. (a) The mouse changes direction (when its velocity is zero) at times 17, 23, and 27.
(b) The mouse is moving most rapidly to the right at time 10 and most rapidly to the left at time 40.
(c) The mouse is farthest to the right when the integral of the velocity, $\int_0^t v(t)\,dt$, is most positive. Since the integral is the sum of the areas above the axis minus the area below the axis, the integral is largest when the velocity is zero at about 17 seconds. The mouse is farthest to the left of center when the integral is most negative at 40 seconds.
(d) The mouse's speed decreases during seconds 10 to 17, from 20 to 23 seconds, and from 24 seconds to 27 seconds.
(e) The mouse is at the center of the tunnel at any time t for which the integral from 0 to t of the velocity is zero. This is true at time 0 and again somewhere around 40 seconds.

CHAPTER 4

Answers for Section 4.1

1. $\dfrac{dy}{dx} = 0$
5. $y' = -12x^{-13}$.
9. $f'(x) = -4x^{-5}$
13. $y' = 18x^2 + 8x - 2$.
17. $\dfrac{dy}{dq} = 8.4q - 0.5$
21. $\dfrac{dy}{dx} = 4ax^3 + 3bx^2 + 2cx + d$
25. (a) $f(x) = -3x + 2, g(x) = 2x + 1$.

$$k(x) = f(x) + g(x)$$
$$= (-3x + 2) + (2x + 1)$$
$$= -x + 3$$
$$k'(x) = -1.$$

Also, $f'(x) = -3, g'(x) = 2$, so $f'(x) + g'(x) = -3 + 2 = -1$.
(b)

$$j(x) = f(x) - g(x)$$
$$= (-3x + 2) - (2x + 1)$$
$$= -5x + 1$$
$$j'(x) = -5.$$

Also, $f'(x) - g'(x) = -3 - 2 = -5$.

29. So far, we can only take the derivative of powers of x and the sums of constant multiples of powers of x. Since we cannot write $\sqrt{x + 3}$ in this form, we cannot yet take its derivative.

33. We cannot write $\dfrac{1}{3x^2+4}$ as the sum of powers of x multiplied by constants.
37. Once again, the x is in the exponent and we haven't learned how to handle that yet.

Answers for Section 4.2

1. (a) $f'(x) = 3x^2 - 2(2x) + 3 + 0 = 3x^2 - 4x + 3$
   (b) $f'(-1) = 3(-1)^2 - 4(-1) + 3 = 10$, $f'(0) = 3(0)^2 - 4(0) + 3 = 3$, $f'(1) = 3(1)^2 - 4(1) + 3 = 2$, $f'(2) = 3(2)^2 - 4(2) + 3 = 7$

(c)

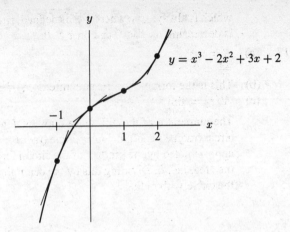

Figure 4.1

The slopes do match the answers we obtained in part (b), because the derivative of a function at a value x is the same as the slope of the tangent line to the graph of the function at the point $(x, f(x))$.

5. $f'(x)$ is $6x^2 - 10x + 3$. We want $f'(1)$, because this is the slope of the tangent line at $x = 1$. $f'(1) = 6(1)^2 - 10(1) + 3 = -1$. We know that the equation of the tangent line at $x = 1$ has the form $y = b + mx$, and since $f'(1) = -1$, we know that $m = -1$. So the tangent line has equation $y = b - x$ for some b. Since the point $(1, f(1))$ is included in the tangent line to $f(x)$ at $x = 1$, we have $f(1) = b - 1$, that is, $-5 = b - 1$, so $b = -4$. The equation of the tangent line is therefore $y = -4 - x$.

9. (a) The yield is $f(5) = 320 + 140(5) - 10(5)^2 = 770$ bushels per acre.
   (b) $f'(x) = 140 - 20x$, so $f'(5) = -100$ bushels per pound of fertilizer. This represents a rate of decrease in the yield of 100 bushels per pound of fertilizer when the number of pounds of fertilizer being used is 5.
   (c) Less should be used, because 5 pounds is on the downslope of the yield curve.

13. (a) $p(x) = x^2 - x$. Now, $p'(x) = 2x - 1 < 0$ when $x < \frac{1}{2}$. So p is decreasing when $x < \frac{1}{2}$.
    (b) $p(x) = x^{\frac{1}{2}} - x$.
    $$p'(x) = \frac{1}{2}x^{-\frac{1}{2}} - 1 < 0$$
    $$\frac{1}{2}x^{-\frac{1}{2}} < 1$$
    $$x^{-\frac{1}{2}} < 2$$
    $$x^{\frac{1}{2}} > \frac{1}{2}$$
    $$x > \frac{1}{4}.$$

    Thus $p(x)$ is decreasing when $x > \frac{1}{4}$.
    (c) $p(x) = x^{-1} - x$.
    $$p'(x) = -1x^{-2} - 1 < 0$$
    $$-x^{-2} < 1$$
    $$x^{-2} > -1,$$

which is always true where x^{-2} is defined, since $x^{-2} = \frac{1}{x^2}$ is always positive. Thus $p(x)$ is decreasing on $x < 0$ and on $x > 0$.

17. (a) $A = \pi r^2$
$\frac{dA}{dr} = 2\pi r$.
(b) This is the formula for the circumference of a circle.
(c) $A'(r) = \lim\limits_{h \to 0} \frac{A(r+h) - A(r)}{h}$
The numerator of the difference quotient denotes the area contained between the inner circle (radius r) and the outer circle (radius $r + h$). As h approaches 0, this area can be approximated by the product of the circumference of the inner circle and the "width" of the area, i.e., h. Dividing this by the denominator, h, we get A' = the circumference of the circle with radius r.

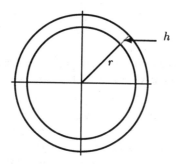

We can also think about the derivative of A as the rate of change of area for a small change in radius. If the radius increases by a tiny amount, the area will increase by a thin ring whose area is simply the circumference at that radius times the small amount. To get the rate of change, we divide by the small amount and obtain the circumference.

Answers for Section 4.3

1. $y' = 10t + 4e^t$.
5. $\frac{dy}{dx} = 3 - 2(\ln 4)4^x$.
9. $P'(t) = Ce^t$.
13. $\frac{dy}{dt} = A\cos t$.
17. $P'(t) = 12.41(\ln 0.94)(0.94)^t$.
21. $f'(t) = \frac{A}{2}t^{(-1/2)} - B\cos t$.
25. (a) $f(x) = 1 - e^x$ crosses the x-axis where $0 = 1 - e^x$, which happens when $e^x = 1$, so $x = 0$. Since $f'(x) = -e^x$, $f'(0) = -e^0 = -1$.
(b) $y = -x$
(c) The negative of the reciprocal of -1 is 1, so the equation of the normal line is $y = x$.
29. (a) $P = 4.1(1 + 0.02)^t = 4.1(1.02)^t$.
(b)
$$\frac{dP}{dt} = 4.1\frac{d}{dt}(1.02)^t = 4.1(1.02)^t(\ln 1.02)$$

$$\left.\frac{dP}{dt}\right|_{t=0} = 4.1(1.02)^0 \ln 1.02 \approx 0.0812$$

$$\left.\frac{dP}{dt}\right|_{t=15} = 4.1(1.02)^{15} \ln 1.02 \approx 0.11.$$

$\frac{dP}{dt}$ is the rate of growth of the world's population; $\left.\frac{dP}{dt}\right|_{t=0}$ and $\left.\frac{dP}{dt}\right|_{t=15}$ are the rates of growth in the years 1975 and 1990, respectively.

33. The tangent lines to $f(x) = \sin x$ have slope $\frac{d}{dx}(\sin x) = \cos x$. The tangent line at $x = 0$ has slope $f'(0) = \cos 0 = 1$ and goes through the point $(0,0)$. Consequently, its equation is $y = g(x) = x$. The approximate value of $\sin \frac{\pi}{6}$ given by this equation is then $g(\frac{\pi}{6}) = \frac{\pi}{6} \approx 0.524$.

Similarly, the tangent line at $x = \frac{\pi}{3}$ has slope $f'(\frac{\pi}{3}) = \cos \frac{\pi}{3} = \frac{1}{2}$ and goes through the point $(\frac{\pi}{3}, \frac{\sqrt{3}}{2})$. Consequently, its equation is $y = h(x) = \frac{1}{2}x + \frac{3\sqrt{3}-\pi}{6}$. The approximate value of $\sin \frac{\pi}{6}$ given by this equation is then $h(\frac{\pi}{6}) = \frac{6\sqrt{3}-\pi}{12} \approx 0.604$.

The actual value of $\sin \frac{\pi}{6}$ is $\frac{1}{2}$, so the approximation from 0 is better than that from $\frac{\pi}{3}$. This is because the slope of the function changes less between $x = 0$ and $x = \frac{\pi}{6}$ than it does between $x = \frac{\pi}{6}$ and $x = \frac{\pi}{3}$.

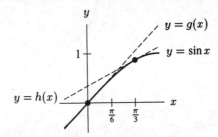

37.

$$g(x) = ax^2 + bx + c \qquad\qquad f(x) = e^x$$
$$g'(x) = 2ax + b \qquad\qquad f'(x) = e^x$$
$$g''(x) = 2a \qquad\qquad f''(x) = e^x$$

So, using $g''(0) = f''(0)$, etc., we have $2a = 1$, $b = 1$, and $c = 1$, and thus $g(x) = \frac{1}{2}x^2 + x + 1$.

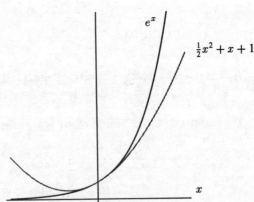

The two functions do look very much alike near $x = 0$. They both increase for large values of x, but e^x increases much more quickly. For very negative values of x, the quadratic goes to ∞ whereas the exponential goes to 0. By choosing a function whose first few derivatives agreed with the exponential when $x = 0$, we got a function which looks like the exponential for x values near 0.

Answers for Section 4.4

1. $f'(x) = 99(x+1)^{98} \cdot 1 = 99(x+1)^{98}$.
5. $w' = 100(\sqrt{t}+1)^{99} \left(\frac{1}{2\sqrt{t}}\right) = \frac{50}{\sqrt{t}}(\sqrt{t}+1)^{99}$.
9. $y' = \dfrac{3s^2}{2\sqrt{s^3+1}}$.
13. $\frac{dy}{dt} = \frac{5}{5t+1}$.
17. $\frac{dy}{dt} = -10\sin 5t$.
21. $f'(t) = \frac{2t}{t^2+1}$.
25. $f'(x) = \cos(3x) \cdot 3 = 3\cos(3x)$.
29. $\frac{dy}{dx} = 2(5+e^x)e^x$.
33. The graph is concave down when $f''(x) < 0$.

$$f'(x) = e^{-x^2}(-2x)$$
$$f''(x) = \left[e^{-x^2}(-2x)\right](-2x) + e^{-x^2}(-2)$$
$$= \frac{4x^2}{e^{x^2}} - \frac{2}{e^{x^2}}$$
$$= \frac{4x^2 - 2}{e^{x^2}} < 0$$

when $4x^2 - 2 < 0$, so $4x^2 < 2$, or $x^2 < \frac{1}{2}$. Hence $-\frac{1}{\sqrt{2}} < x < \frac{1}{\sqrt{2}}$.

37. (a)

$$\frac{dQ}{dt} = \frac{d}{dt}e^{-0.000121t}$$
$$= -0.000121e^{-0.000121t}$$

(b)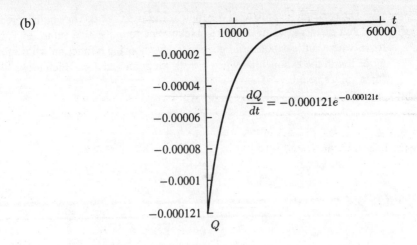

41. (a) $\dfrac{dy}{dt} = -\dfrac{4.9\pi}{6}\sin\left(\dfrac{\pi}{6}t\right)$. It represents the rate of change of the depth of the water.

(b) $\dfrac{dy}{dt}$ is zero where the tangent line to the curve $y(t)$ is horizontal. $\dfrac{dy}{dt} = 0$ occurs when $\sin(\tfrac{\pi}{6}t) = 0$, or at $t =$ 6am, 12 noon, 6pm and 12 midnight. When $\dfrac{dy}{dt} = 0$, the depth of the water is no longer changing. Therefore, it has either just finished rising or just finished falling, and we know that the harbor's level is at a maximum or a minimum.

Answers for Section 4.5

1. By the product rule,
$$f'(x) = 2x(x^3+5) + x^2(3x^2) = 2x^4 + 3x^4 + 10x = 5x^4 + 10x.$$

Alternatively,
$$f'(x) = (x^5 + 5x^2)' = 5x^4 + 10x.$$

The two answers should, and do, match.

5. $y' = 2^x + x(\ln 2)2^x = 2^x(1 + x\ln 2)$.
9. $y' = (3t^2 - 14t)e^t + (t^3 - 7t^2 + 1)e^t = (t^3 - 4t^2 - 14t + 1)e^t$.
13. $y' = 1 \cdot e^{-t^2} + te^{-t^2}(-2t)$
17.
$$\begin{aligned} f'(w) &= (e^{w^2})(10w) + (5w^2 + 3)(e^{w^2})(2w) \\ &= 2we^{w^2}(5 + 5w^2 + 3) \\ &= 2we^{w^2}(5w^2 + 8). \end{aligned}$$

21. $f(x) = x^2 e^{-x}$, $f(0) = 0$
$f'(x) = 2xe^{-x} + x^2 e^{-x} \cdot (-1) = e^{-x}(2x - x^2)$, so $f'(0) = 0$. Tangent line is $y = 0$ (the x-axis).

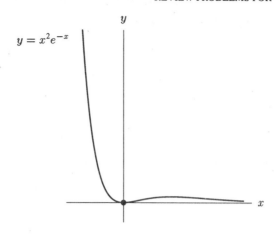

Figure 4.2

25. (a) $G'(z) = F'(z)H(z) + H'(z)F(z)$, so
$G'(3) = F'(3)H(3) + H'(3)F(3) = 4 \cdot 1 + 3 \cdot 5 = 19$.
(b) $G'(w) = \dfrac{F'(w)H(w) - H'(w)F(w)}{[H(w)]^2}$, so $G'(3) = \dfrac{4(1)-3(5)}{1^2} = -11$.

REVIEW PROBLEMS FOR CHAPTER FOUR

1. $f'(x) = 3x^2 - 6x + 5$
5. $f'(t) = \dfrac{d}{dt}\left(2te^t - \dfrac{1}{\sqrt{t}}\right) = 2e^t + 2e^t t + \dfrac{1}{2t^{3/2}}$.
9. $\dfrac{dy}{dx} = e^{3x} + x \cdot 3e^{3x} = e^{3x}(1 + 3x)$
13. $g'(x) = \dfrac{d}{dx}\left(x^{\frac{1}{2}} + x^{-1} + x^{-\frac{3}{2}}\right) = \dfrac{1}{2}x^{-\frac{1}{2}} - x^{-2} - \dfrac{3}{2}x^{-\frac{5}{2}}$.
17. $h'(r) = 5 \cdot 2(\sin r)(\cos r) = 10 \sin r \cos r$
21. (a) If the distance $s(t) = 20e^{\frac{t}{2}}$, then the velocity, $v(t)$, is given by
$$v(t) = s'(t) = \left(20e^{\frac{t}{2}}\right)' = \left(\dfrac{1}{2}\right)\left(20e^{\frac{t}{2}}\right) = 10e^{\frac{t}{2}}.$$
(b) Observing the differentiation in (a), we note that
$$s'(t) = v(t) = \dfrac{1}{2}\left(20e^{\frac{t}{2}}\right).$$
Substituting $s(t)$ for $20e^{\frac{t}{2}}$, we obtain $s'(t) = \dfrac{1}{2}s(t)$.
25. Since we're given that the instantaneous rate of change of T at $t = 30$ is 2, we want to choose a and b so that the derivative of T agrees with this value. Differentiating, $T'(t) = ab \cdot e^{-bt}$. Then we have
$$2 = T'(30) = abe^{-30b} \text{ or } e^{-30b} = \dfrac{2}{ab}$$

We also know that at $t = 30, T = 120$, so

$$120 = T(30) = 200 - ae^{-30b} \text{ or } e^{-30b} = \frac{80}{a}$$

Thus $\frac{80}{a} = e^{-30b} = \frac{2}{ab}$, so $b = \frac{1}{40} = 0.025$ and $a = 169.36$.

CHAPTER 5

Answers for Section 5.1

1.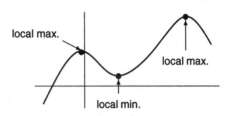

   Figure 5.1

5. $f'(x) = 12x^3 - 12x^2$. To find critical points, we set $f'(x) = 0$. This implies $12x^2(x-1) = 0$. So the critical points of f are $x = 0$ and $x = 1$. To the left of $x = 0$, $f'(x) < 0$. Between $x = 0$ and $x = 1$, $f'(x) < 0$. To the right of $x = 1$, $f'(x) > 0$. Therefore, $f(1)$ is a local minimum, but $f(0)$ is not a local extremum.

   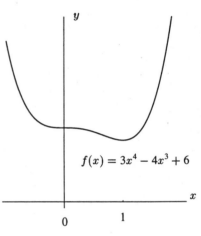

9. $f'(x) = 4xe^{5x} + 2x^2 e^{5x} \cdot 5 = 2e^{5x}x(5x+2)$. Notice that $e^{5x} > 0$ for all x. So, the critical points are $x = 0$ and $x = -2/5$.
   To the left of $x = -2/5$, $f'(x) > 0$.
   Between $x = -2/5$ and $x = 0$, $f'(x) < 0$.
   To the right of $x = 0$, $f'(x) > 0$.
   So, $f(-2/5)$ is a local maximum, $f(0)$ a local minimum. Notice that in the figure, you can barely discern the local maximum and minimum.

   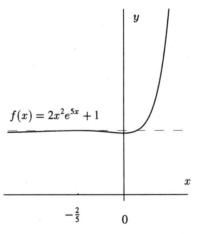

13. $f(x) = x - \ln x$, where $0.1 \le x \le 2$.
    (a) $f'(x) = 1 - \frac{1}{x}$. This is zero only when $x = 1$, and so $x = 1$ is the only critical point of f. Now $f'(x)$ is positive when $x > 1$, and negative when $x < 1$. Thus $f(1) = 1$ is a local minimum.

(b) We have, by looking at the endpoints and the critical point,

$$f(0.1) = 0.1 - \ln(0.1) \approx 2.4026$$
$$f(1) = 1$$
$$f(2) = 2 - \ln 2 \approx 1.3069.$$

Thus $x = 0.1$ gives the global maximum and $x = 1$ gives the global minimum.

17. Using the product rule on the function $f(x) = axe^{bx}$, we have $f'(x) = ae^{bx} + abxe^{bx}$. We want $f(\frac{1}{3}) = 1$, and since this is to be a maximum, we require $f'(\frac{1}{3}) = 0$. These conditions give

$$f\left(\frac{1}{3}\right) = a\left(\frac{1}{3}\right)e^{\left(\frac{1}{3}\right)b} = 1,$$
$$f'\left(\frac{1}{3}\right) = ae^{\left(\frac{1}{3}\right)b} + ab\left(\frac{1}{3}\right)e^{\left(\frac{1}{3}\right)b} = 0.$$

Since $ae^{(\frac{1}{3})b}$ is non-zero, we can divide both sides of the second equation by $ae^{(\frac{1}{3})b}$ to obtain $0 = 1 + \frac{b}{3}$. This implies $b = -3$. Plugging $b = -3$ into the first equation gives us $a(\frac{1}{3})e^{-1} = 1$, or $a = 3e$. How do we know we have a maximum at $x = \frac{1}{3}$ and not a minimum? Since $f'(x) = ae^{bx}(1 + bx) = (3e)e^{-3x}(1 - 3x)$, and $(3e)e^{-3x}$ is always positive, it follows that $f'(x) > 0$ when $x < \frac{1}{3}$ and $f'(x) < 0$ when $x > \frac{1}{3}$. Since f' is positive to the left of $x = \frac{1}{3}$ and negative to the right of $x = \frac{1}{3}$, $f(\frac{1}{3})$ is a local maximum.

21.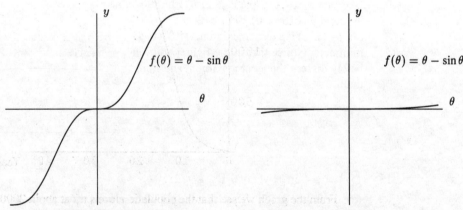

Figure 5.2: Graph of $f(\theta)$ **Figure 5.3**: Graph of $f(\theta)$ Zoomed In

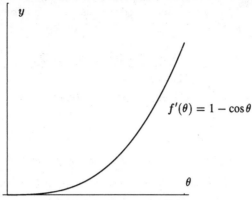

Figure 5.4: Graph of $f'(\theta)$

(a) In Figure 5.2, we see that $f(\theta) = \theta - \sin\theta$ definitely has a zero at $\theta = 0$. To see if it has any other zeros near the origin, we use our calculator to zoom in. (See Figure 5.3.) No extra root seems to appear no matter how close to the origin we zoom. However, zooming can never tell you for sure that there is not a root that you have not found yet.

(b) Using the derivative we can argue that there is no other zero. $f'(\theta) = 1 - \cos\theta$. Since $\cos\theta < 1$ for $0 < \theta \leq 1$, $f'(\theta) > 0$ for $0 < \theta \leq 1$. Thus, f increases for $0 < \theta \leq 1$. Consequently, we conclude that the only zero of f is the one at the origin. If f had another zero at x_0, $x_0 > 0$, f would have to "turn around", and recross the x-axis at x_0. But if this were the case, f' would be nonpositive somewhere, which we know to be impossible.

25. (a)
$$P(t) = \frac{2000}{1 + e^{(5.3 - 0.4t)}}$$

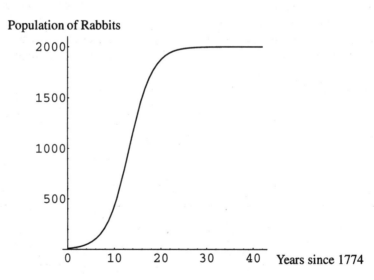

From the graph we see that the population levels off at about 2000 rabbits.

(b) The population appears to have been growing fastest when there were about 500 rabbits, just over 10 years after Captain Cook left them there.

(c) The rabbits reproduce quickly, so their population initially grew very rapidly. Limited food and space availability or perhaps predators on the island probably accounts for the population being unable to grow past 2000.

Answers for Section 5.2

1.

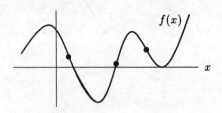

Figure 5.5: 3 inflection points

5. We first find the maxima and minima of $f(x) = x^4 - 4x^3 + 10$. Since $f'(x) = 4x^3 - 12x^2$, so setting the derivative equal to 0 and factoring yields

$$4x^3 - 12x^2 = 0$$
$$4x^2(x - 3) = 0$$

So $x = 0$ and $x = 3$ are the critical points of $f(x)$. Using the first derivative test,

$$f'(x) < 0 \text{ for } x < 0$$
$$f'(x) < 0 \text{ for } 0 < x < 3 \quad \text{and}$$
$$f'(x) > 0 \text{ for } x > 3.$$

From this we conclude that $f(x)$ does not have a local minimum or local maximum at $x = 0$, and that $f(x)$ has a local minimum at $x = 3$. We now find points of inflection of $f(x)$.

$$f''(x) = 12x^2 - 24x.$$

Setting the second derivative equal to 0 and factoring yields

$$12x^2 - 24x = 0$$
$$12x(x - 2) = 0$$

So $x = 0$ and $x = 2$ may be points of inflection of $f(x)$. Using the second derivative test,

$$f''(x) > 0 \text{ for } x < 0$$
$$f''(x) < 0 \text{ for } 0 < x < 2 \quad \text{and}$$
$$f''(x) > 0 \text{ for } x > 2.$$

Since $f''(x)$ changes sign at both $x = 0$ and $x = 2$, both are points of inflection for $f(x)$. Furthermore, $f(x)$ is concave down on the interval $0 < x < 2$ and concave up elsewhere. Using all of this information, we can now sketch the graph, as follows:

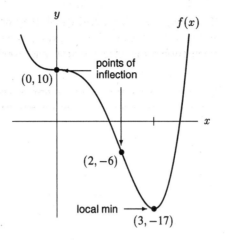

Figure 5.6

9.

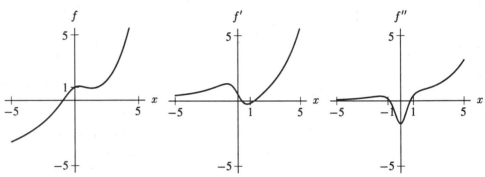

Figure 5.7

There is no way to do this algebraically since there is no way to solve the necessary equations: $f(x) = 0$, $f'(x) = 0$, and $f''(x) = 0$. To find the maxima, minima, and inflection points, we must use a calculator to approximate the solutions of these equations. We have

$$f'(x) = e^{\frac{x}{2}}\left(\frac{1}{2}\right) - \frac{1}{x^2 + 1}(2x) = \frac{1}{2}e^{\frac{x}{2}} - \frac{2x}{x^2 + 1},$$

$$f''(x) = \frac{1}{2}e^{\frac{x}{2}}\left(\frac{1}{2}\right) - \frac{(x^2 + 1)(2) - 2x(2x)}{(x^2 + 1)^2} = \frac{1}{4}e^{\frac{x}{2}} + \frac{2(x^2 - 1)}{(x^2 + 1)^2}$$

By looking at the graph of f closely with a calculator, we find that the intercepts are exactly $(0, 1)$ and approximately $(-0.934, 0)$. To find the local maxima and minima, we look for zeros on the graph of f'. There are zeros at 0.325 and 1.313. Plugging these values into f, we get a local maximum at $(0.325, 1.076)$ and a local minimum at $(1.313, 0.926)$. These are, however, only local, since $f(x) \to \pm\infty$ as $x \to \pm\infty$. To find inflection points, we look for zeros of f''. Again, by the graphing calculator, there are zeros (and thus inflection points judging from graphs of f and f'') at $x = 0.747$ and $x = -0.867$. The coordinates of the curve at the inflection points are $(0.747, 1.009)$ and $(-0.867, 0.088)$.

13. (a)

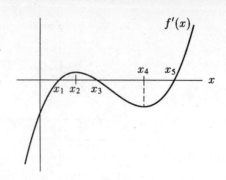

(b) $f'(x)$ changes sign at x_1, x_3, and x_5.
(c) $f'(x)$ has local extrema at x_2 and x_4.

17.

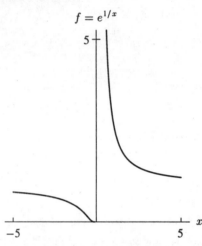

(a) $f'(x) = (\frac{-1}{x^2})(e^{\frac{1}{x}})$. Thus $f'(x) < 0$ for all $x \neq 0$, which means $f(x)$ is decreasing everywhere it is defined.

(b) $f''(x) = \frac{1}{x^4}e^{\frac{1}{x}} + \frac{2}{x^3}e^{\frac{1}{x}} = \frac{(2x+1)}{x^4}e^{\frac{1}{x}}$.
$f''(x) = 0$ when $x = -1/2$.
$f''(x) < 0$ for $x < -1/2$ and $f''(x) > 0$ for $-1/2 < x < 0$ and $x > 0$.
So, $f(x)$ is concave up for $x > 0$ and $-1/2 < x < 0$, and concave down for $x < -1/2$.

21.

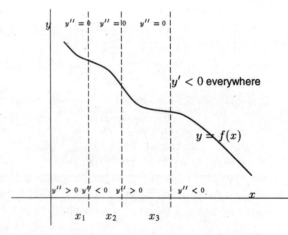

25.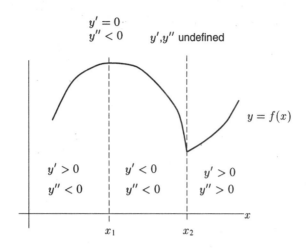

Answers for Section 5.3

1. The parameter b simply shifts the parabola vertically without changing its shape. The parameter a shifts the vertex of parabola both vertically and horizontally. If $y = x^2 + ax + b$, $y' = 2x + a$, and the parabola vertex is at the critical point $-\frac{a}{2}$, so the vertex shifts to $(-\frac{a}{2}, -\frac{a^2}{4} + b)$.

5. A affects the amplitude (e.g. height) of the curve. B affects the frequency. C causes a phase shift to the left or right.

   Changing A:

   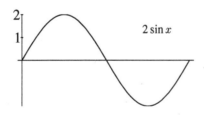 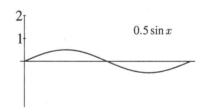

   Changing B:

   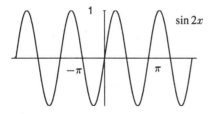 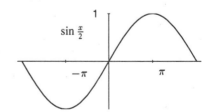

Changing C:

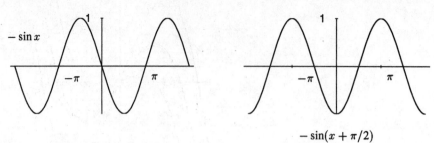

9. $y' = a(1 - \frac{b}{x})$, so the critical point is at $x = b$.

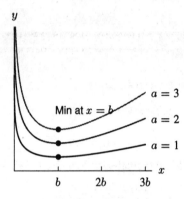

Figure 5.8: varying a

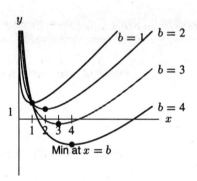

Figure 5.9: varying b

13. (a) We have $p'(x) = 3x^2 - a$, and (see solution to 12)

$$\begin{array}{c|c|c|c} p \text{ increasing} & & p \text{ decreasing} & & p \text{ increasing} \\ \hline & -\sqrt{\frac{a}{3}} & & \sqrt{\frac{a}{3}} & \end{array}$$

Local maximum: $p(-\sqrt{\frac{a}{3}}) = \frac{-a\sqrt{a}}{\sqrt{27}} + \frac{a\sqrt{a}}{\sqrt{3}} = +\frac{2a\sqrt{a}}{3\sqrt{3}}$

Local minimum: $p(\sqrt{\frac{a}{3}}) = -p(-\sqrt{\frac{a}{3}}) = -\frac{2a\sqrt{a}}{3\sqrt{3}}$

(b) Increasing the value of a moves the critical points of p away from the y-axis, and moves the critical values away from the x-axis. Thus, the "bumps" get higher and further apart. At the same time, increasing the value of a spreads the zeros of p further apart (while leaving the one at the origin fixed).

(c)

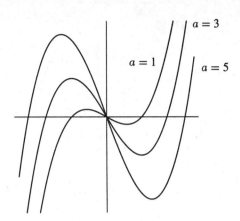

Answers for Section 5.4

1. First find marginal revenue and marginal cost.

$$MR = R'(q) = 450$$

$$MC = C'(q) = 6q$$

Setting $MR = MC$ yields $6q = 450$, marginal cost is equal to marginal revenue when

$$q = \frac{450}{6} = 75 \text{ units.}$$

Is profit maximized at $q = 75$? Profit $= R(q) - C(q)$

$$R(75) - C(75) = 450(75) - (10000 + 3(75)^2)$$
$$= 33{,}750 - 26{,}875 = \$6875$$

Testing $q = 74$ and $q = 76$

$$R(74) - C(74) = 450(74) - (10000 + 3(74)^2)$$
$$= 33{,}000 - 26{,}428 = \$6872$$

$$R(76) - C(76) = 450(76) - (10000 + 3(76)^2)$$
$$= 34{,}200 - 27{,}328 = \$6872$$

Since profit at $q = 75$ is more than profit at $q = 74$ and $q = 76$, we conclude that profit is maximized locally at $q = 75$. The only endpoint we need to check is $q = 0$.

$$R(0) - C(0) = 450(0) - (10000 + 3(0)^2)$$
$$= -\$10{,}000$$

This is clearly not a maximum, so we conclude that the profit is maximized globally at $q = 75$, and the total profit at this production level is $\$6{,}875$.

5.

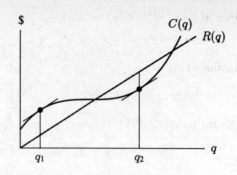

Figure 5.10

We find q_1 and q_2 by checking to see where the slope of the tangent line of $C(q)$ is equal to the slope of the tangent line of $R(q)$. Also, q_1 is a local maximum, since $C(q) - R(q)$ is greatest at this point, and q_2 is a local minimum, since $R(q) - C(q)$ is greatest at this point.

9. (a)
$$C'(q) = \frac{K}{a}q^{(1/a)-1}, \quad C''(q) = \frac{K}{a}\left(\frac{1}{a}-1\right)q^{(1/a)-2}.$$

If $a > 1$, $C''(q) < 0$, so C is concave down.

(b)
$$a(q) = \frac{C(q)}{q} = \frac{Kq^{1/a} + F}{q}$$
$$M(q) = \frac{K}{a}q^{1/a-1}$$

so $a(q) = M(q)$ means
$$\frac{Kq^{1/a} + F}{q} = \frac{K}{a}q^{(1/a)-1}.$$

Solving,
$$Kq^{1/a} + F = \frac{K}{a}q^{1/a}$$
$$K\left(\frac{1}{a}-1\right)q^{1/a} = F$$
$$q = \left[\frac{Fa}{K(1-a)}\right]^a.$$

13. Let x equal the number of chairs ordered in excess of 300, so $0 \leq x \leq 100$.

$$\text{Revenue} = R = (90 - 0.25x)(300 + x)$$
$$= 27{,}000 - 75x + 90x - 0.25x^2 = 27{,}000 + 15x - 0.25x^2$$

At a critical point $dR/dx = 0$. $dR/dx = 15 - 0.5x$ so $x = 30$ gives a maximum revenue of $\$27{,}225$ since the graph of R is a parabola which opens downwards. The minimum is $\$0$ (when no chairs are sold).

17. Since $a(q) = \frac{C(q)}{q}$, $C(q) = a(q) \cdot q$. Thus $C'(q) = q \cdot a'(q) + a(q)$, and so $C'(q_0) = q_0 \cdot a'(q_0) + a(q_0)$.

    Since t_1 is the line tangent to $a(q)$ at $q = q_0$, the slope of the line t_1 is $a'(q_0)$, and the equation of the line is

    $$y = a(q_0) + a'(q_0) \cdot (q - q_0) = a'(q_0) \cdot q + \big(a(q_0) - a'(q_0) \cdot q_0\big).$$

    Thus the y-intercept of t_1 is given by $a(q_0) - q_0 \cdot a'(q_0)$, and the equation of the line t_2 is

    $$y = 2 \cdot a'(q_0) \cdot q + \big(a(q_0) - a'(q_0) \cdot q_0\big)$$

    since t_2 has twice the slope of t_1. Let's compute y for $q = q_0$:

    $$y = 2 \cdot a'(q_0) \cdot q_0 + \big(a(q_0) - a'(q_0) \cdot q_0\big) = q_0 \cdot a'(q_0) + a(q_0) = C'(q_0).$$

    Hence $C'(q_0)$ is given by the point on t_2 where $q = q_0$.

    (This rule also would work to solve Problem 15. In that case, t_1 coincides with $a(q)$, and so $C'(q)$ is the line with the same y–intercept and twice the slope as t_1.)

Answers for Section 5.5

1. (a) If the price of yams is \$2/pound, the quantity sold will be

    $$q = 5000 - 10(2)^2 = 5000 - 40 = 4960$$

    so 4960 pounds will be sold.

    (b) Elasticity of demand is given by

    $$E = \frac{p}{q} \cdot \frac{dq}{dp} = \frac{p}{q} \cdot \frac{d}{dp}[5000 - 10p^2] = \frac{p}{q} \cdot (-20p) = \frac{-20p^2}{q}$$

    Substituting $p = 2$ and $q = 4960$ yields

    $$E = \frac{-20(2)^2}{4960} = \frac{-80}{4960} = -.016$$

    Since $|E| < 1$ the demand is inelastic, so it would be more correct to say "People must have their yams and will buy them no matter what the price."

5. Demand for high-definition TV's will be elastic, since it is not a necessary item. If the prices are too high, people will not choose to buy them, so price changes will cause demand changes.

9. Since $R = pq$, we have $dR/dp = p(dq/dp) + q$. We are assuming that

    $$E = (p/q)dq/dp < -1$$

    Multiplication by q gives

    $$p(dq/dp) < -q$$

    and hence

    $$dR/dp = p(dq/dp) + q < 0$$

REVIEW PROBLEMS FOR CHAPTER FIVE

1. (a) We wish to investigate the behavior of $f(x) = x^3 - 3x^2$ on the interval $-1 \leq x \leq 3$. We find:

$$f'(x) = 3x^2 - 6x = 3x(x - 2)$$
$$f''(x) = 6x - 6 = 6(x - 1)$$

   (b) The critical points of f are $x = 2, 0$, since $f'(x) = 0$ here. Using the second derivative test, we find that $x = 0$ is a local maximum since $f'(0) = 0$ and $f''(0) = -6 < 0$, and that $x = 2$ is a local minimum since $f'(2) = 0$ and $f''(2) = 6 > 0$.

   (c) There is an inflection point at $x = 1$ since f'' changes sign at $x = 1$.

   (d) At the critical points, $f(0) = 0$ and $f(2) = -4$.
   At the endpoints: $f(-1) = -4, f(3) = 0$.
   So the global maxima are $f(0) = 0$ and $f(3) = 0$, while the global minima are $f(-1) = -4$ and $f(2) = -4$.

   (e)

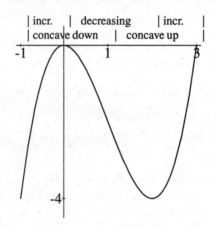

5.
$$\lim_{x \to \infty} \ln(x^2 + 1) = \infty, \text{ and}$$

$$\lim_{x \to -\infty} \ln(x^2 + 1) = \infty,$$

   (a) Using the chain rule,

$$f'(x) = \frac{1}{x^2 + 1} \cdot 2x = \frac{2x}{x^2 + 1}$$

   and using the quotient rule,

$$f''(x) = \frac{2 \cdot (x^2 + 1) - 2x(2x)}{(x^2 + 1)^2} = \frac{2x^2 + 2 - 4x^2}{(x^2 + 1)^2}$$
$$= \frac{-2x^2 + 2}{(x^2 + 1)^2}$$

(b) To find the critical points of $f(x)$, we let $f'(x) = 0$ and solve for x.

$$\frac{2x}{x^2 + 1} = 0$$

This equation is satisfied only if $x = 0$, so $x = 0$ is the only critical point of $f(x)$.

(c) To find the inflection points of $f(x)$, we let $f''(x) = 0$ and solve for x.

$$\frac{-2x^2 + 2}{(x^2 + 1)^2} = 0$$

This is satisfied only if

$$-2x^2 + 2 = 0$$
$$2x^2 = 2$$
$$x^2 = 1$$

So $f''(x) = 0$ if $x = 1$ or $x = -1$. Note that

$$f''(x) < 0 \text{ for } x < -1$$
$$f''(x) > 0 \text{ for } -1 < x < 1, \text{ and}$$
$$f''(x) < 0 \text{ for } x > 1$$

So $f(x)$ has inflection points at $x = -1$ and $x = 1$.

(d) The critical point is $x = 0$.

$$f(0) = \ln(0^2 + 1) = \ln 1 = 0$$

(e) Using the first derivative test

$$f'(x) < 0 \text{ for } x < 0, \text{ and}$$
$$f'(x) > 0 \text{ for } x > 0$$

So we conclude that $f(x)$ has a local minimum at $x = 0$. Since $x = 0$ is the only critical point, and since $\lim_{x \to -\infty} \ln(x^2 + 1) = \lim_{x \to \infty} \ln(x^2 + 1) = \infty$, we conclude that $f(x)$ has a global minimum at $x = 0$, and no global maximum.

(f)

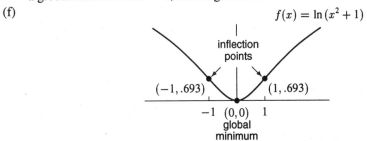

Figure 5.11

f is decreasing for $x < 0$ and increasing for $x > 0$. $f(x)$ is concave down for $x < -1$ and for $x > 1$, and is concave up for $-1 < x < 1$.

9. $\lim_{x \to +\infty} f(x) = +\infty$, and $\lim_{x \to -\infty} f(x) = -\infty$.
   There are no asymptotes.
   $f'(x) = 5x^4 - 45x^2 = 5x^2(x^2 - 9) = 5x^2(x+3)(x-3)$.
   The critical points are $x = 0$, $x = \pm 3$. f' changes sign at 3 and -3 but not at 0.
   $f''(x) = 20x^3 - 90x = 10x(2x^2 - 9)$. f'' changes sign at $0, \pm 3/\sqrt{2}$.
   So, inflection points are at $x = 0$, $x = \pm \frac{3}{\sqrt{2}}$.

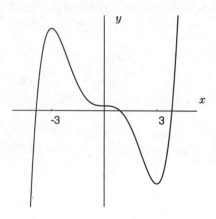

TABLE 5.1

x		-3		$\frac{-3}{\sqrt{2}}$		0		$\frac{3}{\sqrt{2}}$		3	
f'	$+$	0	$-$		$-$	0	$-$		$-$	0	$+$
f''	$-$		$-$	0	$+$	0	$-$	0	$+$		$+$
f	↗⌢		↘⌢		↘⌣		↘⌢		↘⌣		↗⌣

Thus, $f(-3)$ is a local maximum; $f(3)$ is a local minimum. There are no global maxima or minima.

13. $\lim_{x \to +\infty} f(x) = \lim_{x \to -\infty} f(x) = 1$.
    Thus, $y = 1$ is a horizontal asymptote. Since $x^2 + 1$ is never 0, there are no vertical asymptotes.

$$f'(x) = \frac{2x(x^2+1) - x^2(2x)}{(x^2+1)^2} = \frac{2x}{(x^2+1)^2}.$$

So, $x = 0$ is the only critical point.

$$f''(x) = \frac{2(x^2+1)^2 - 2x \cdot 2(x^2+1) \cdot 2x}{(x^2+1)^4}$$
$$= \frac{2(x^2+1-4x^2)}{(x^2+1)^3}$$
$$= \frac{2(1-3x^2)}{(x^2+1)^3}.$$

So, $x = \pm \frac{1}{\sqrt{3}}$ are inflection points.

TABLE 5.2

x		$\frac{-1}{\sqrt{3}}$		0		$\frac{1}{\sqrt{3}}$	
f'	−		−	0	+		+
f''	−	0	+		+	0	−
f	↘⌢		↘⌣		↗⌣		↗⌢

Thus, $f(0) = 0$ is a local and global minimum.

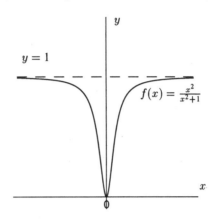

17. (a) $(-\infty, -1)$ decreasing, $(-1, 0)$ increasing, $(0, 1)$ decreasing, $(1, \infty)$ increasing.
    (b) local minima at $f(-1)$ and $f(1)$, local maximum at $f(0)$.

21. (a) To maximize benefit (surviving young), we pick 10, because that's the highest point of the benefit graph.
    (b) To optimize (the vertical distance between the curves) we can either do it by inspection or note that the slopes of the two curves will be the same where the difference is maximized. Either way, one gets approximately 9.

25. To maximize revenue, we first must find an expression for revenue in terms of price. We know that $R(p) = pq$, where p=price and q=quantity sold. We now need to find an expression for q in terms of p. Using the information given, we find that

$$q = 4000 + \frac{(4.00 - p)}{.25}(200)$$

Simplification of q yields

$$q = 4000 + 800(4 - p)$$
$$= 4000 + 3200 - 800p$$
$$= 7200 - 800p$$

We can now get an expression for revenue in terms of price.

$$R(p) = qp = (7200 - 800p)p$$
$$= 7200p - 800p^2$$

We want to maximize this function in terms of p. First find the critical points by finding the derivative.

$$R'(p) = 7200 - 1600p$$

Setting $R'(p) = 0$ and solving for p yields

$$7200 - 1600p = 0$$
$$1600p = 7200$$
$$p = 4.5$$

Since $R'(p) > 0$ for $p < 4.5$ and $R'(p) < 0$ for $p > 4.5$, we conclude that revenue has a local maximum at $p = 4.5$. Since this is the only critical point, we conclude that it is the global maximum. So revenue is maximized at a price of \$4.50. The quantity sold at this amount is given by

$$q = 7200 - 800(4.50) = 3600$$

and the total revenue is

$$R(4.5) = 7200(4.5) - 800(4.5)^2 = \$16{,}200.$$

29. (a) If the water is colder, the air warms it. If the water is warmer, the air cools it.
    (b)
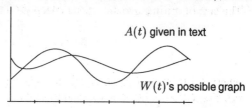

(c) When the graphs intersect, the water temperature is not changing (since it is exactly the same as the air temperature). In other words, the derivative of the water temperature will be zero. Thus, the extrema of the water temperature occur at these intersections.
(d) The greater $A(t) - W(t)$ is, the greater the rate of change of the water temperature.
(e) When $A(t) - W(t)$ reaches a maximum or minimum, the rate at which $W(t)$ changes reaches a maximum or minimum. In other words, $W'(t)$ is at a maximum or minimum and therefore $W''(t) = 0$ and $W'''(t)$ changes sign. Thus $W(t)$ has inflection points whenever $A(t) - W(t)$ reaches a maximum or minimum.
(f)
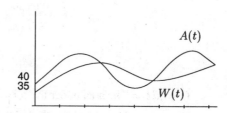

CHAPTER 6

Answers for Section 6.1

1. (a) Since $f(x)$ is positive on the interval from 0 to 6, the integral is equal to the area under the curve. By examining the graph, we can measure and see that the area under the curve is 20 square units, so
$$\int_0^6 f(x)dx = 20.$$

   (b) The average value of $f(x)$ on the interval from 0 to 6 equals the definite integral we calculated in part (a) divided by the size of the interval. Thus
$$\text{Average Value} = \frac{1}{6}\int_0^6 f(x)dx = 3\frac{1}{3}.$$

5. The cost of drilling a well x meters deep is $C(x)$ with
$$C(x) = \int_0^x C'(x)dx + C(0)$$

where $C(0) = 1,000,000$ riyals, the fixed cost. To determine $\int_0^x C'(x)dx$, we graph the function $C'(x) = 4000 + 10x$ and look at the area under the curve in Figure 6.1.

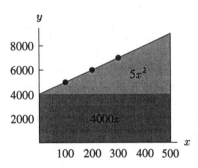

Figure 6.1

$C'(x)dx$ = The area under the curve between 0 and x
  = the area of the shaded rectangle + the area of the shaded triangle
  $= 4000x + x(10x)\left(\dfrac{1}{2}\right)$
  $= 4000x + 5x^2$

Thus
$$C(x) = 5x^2 + 4000x + 1,000,000.$$

9. (a) Using the Fundamental Theorem we get that the cost of producing thirty bicycles is

$$C(30) = \int_0^{30} C'(q)\,dq + C(0)$$

or

$$\int_0^{30} \frac{600}{0.3q+5}\,dq + \$2000 = \$4059.24$$

(b) If the bikes are sold for \$200 each the total revenue for producing 30 bicycles is

$$30 \cdot 200 = \$6000$$

so the total profit is

$$\$6000 - \$4059.24 = \$1950.76$$

(c) The marginal profit on the 31st bicycle is the difference between the marginal cost of producing the 31st bicycle and the profit. Thus the marginal profit is

$$C'(30) - 200 = 200 - \frac{600}{14} = \$157.14$$

13. Taking left-hand sums we get

$$5 \cdot (10.82 + 13.06 + 14.61 + 14.99 + 18.60 + 19.33) = 5 \cdot 91.41 = 457.05.$$

Taking left-hand sums we get

$$5 \cdot (13.06 + 14.61 + 14.99 + 18.60 + 19.33 + 22.46) = 5 \cdot 103.05 = 515.25.$$

Taking the average we get that 486.15 quadrillion BTU is produced between the years 1960 and 1990 in the United States.

Answers for Section 6.2

1. (a) The area under the curve is greater for species B for the first 5 years. Thus species B has a larger population after 5 years. After 10 years, the area under the graph for species B is still greater so species B has a greater population after 10 years as well.
   (b) Unless something happens that we cannot predict now, species A will have a larger population after 20 years. It looks like species A will continue to quickly increase, while species B will add only a few new plants each year.

5. Looking at Figure 6.24 we note that Product B has a greater peak concentration than Product A; Product A peaks sooner than Product B; Product B has a greater overall bioavailability than Product A. Since we are looking for the product providing the faster response, Product A should be used as it peaks sooner.

Answers for Section 6.3

1.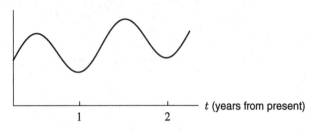

   Figure 6.2

   The graph reaches a peak each summer, and a trough each winter. The graph shows sunscreen sales increasing from cycle to cycle. This gradual increase may be due in part to inflation and to population growth.

5. (a) Solve for $P(t) = P$.

$$100000 = \int_0^{10} Pe^{0.10(10-t)}dt = Pe\int_0^{10} e^{-0.10t}dt$$

$$= \frac{Pe}{-0.10}e^{-0.10t}\Big|_0^{10} = Pe(-3.678 + 10)$$

$$= P \cdot 17.183$$

   So, $P \approx \$5820$ per year.

   (b) To answer this, we'll calculate the present value of $100,000:

$$100000 = Pe^{0.10(10)}$$

$$P \approx \$36{,}787.94.$$

9. One good way to approach the problem is in terms of present values. In 1980, the present value of Germany's loan was 20 billion DM. Now let's figure out the rate that the Soviet Union would have to give money to Germany to pay off 10% interest on the loan by using the formula for the present value of a continuous stream. Since the Soviet Union sends gas at a constant rate, the rate of deposit, $P(t)$, is a constant c. Since they don't start sending the gas until after 5 years have passed, the present value of the loan is given by:

$$\text{Present Value} = \int_5^\infty P(t)e^{-rt}\,dt.$$

   We want to find c so that

$$20{,}000{,}000{,}000 = \int_5^\infty ce^{-rt}\,dt = c\int_5^\infty e^{-rt}\,dt$$

$$= c\lim_{b\to\infty}(-10e^{-0.10t})\Big|_5^b = ce^{-0.10(5)} \approx 6.065c.$$

   Dividing, we see that c should be about 3.3 billion DM per year. At 0.10 DM per m^3 of natural gas, the Soviet Union must deliver gas at the constant, continuous rate of about 33 billion m^3 per year.

Answers for Section 6.4

1. (a) Looking at the Figure 6.36 we see that the equilibrium price is roughly $30 giving an equilibrium quantity of 125 units.

   (b) Consumer surplus is the area above p^* and below the demand curve. Graphically this is represented by the shaded area in Figure 6.3.

   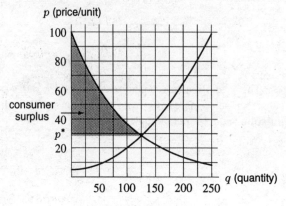

   Figure 6.3

   From the graph we can estimate the shaded area to be roughly 14 squares where each square represents ($25/unit)·(10 units). Thus the consumer surplus is

   $$14 \cdot \$250 = \$3500$$

   (c) Producer surplus is the area under p^* and above the supply curve. Graphically this is represented by the shaded area in Figure 6.3.

   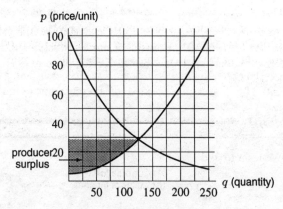

   Figure 6.4

   From the graph we can estimate the shaded area to be roughly 8 squares where each square represents ($25/unit)·(10 units). Thus the producer surplus is

   $$8 \cdot \$250 = \$2000$$

(d)
$$\text{Total gains from trade} = \text{Consumer surplus} + \text{producer surplus}$$
$$= \$3500 + \$2000$$
$$= \$5500$$

5.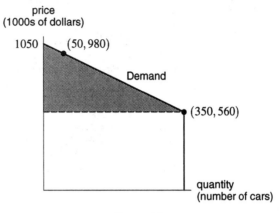

Figure 6.5

Measuring money in thousands of dollars, the equation of the line representing the demand curve passes through (50, 980) and (350, 560). So the equation is $y - 560 = \frac{420}{-300}(x - 350)$, i.e. $y - 560 = -\frac{7}{5}x + 1050$.
The consumer surplus is thus

$$\int_0^{350} \left(-\frac{7}{5}x + 1050\right) dx - (350)(560) = -\frac{7}{10}x^2 + 1050x \bigg|_0^{350} - 196000$$
$$= 85750.$$

(Note that $85750 = \frac{1}{2} \cdot 490 \cdot 350$, the area of the triangle in the diagram. We thus could have avoided the formula for consumer surplus in solving the problem.)
Recalling that our unit measure for the price axis is \$1000/car, the consumer surplus is \$85,750,000.

Answers for Section 6.5

1.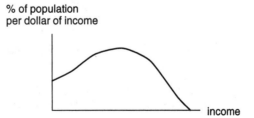

Figure 6.6: Density function

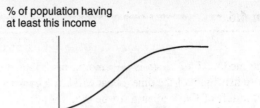

Figure 6.7: Cumulative distribution function

5. (a) The first item is sold at the point at which the graph is first greater than zero. Thus the first item is sold at $t = 30$ or January 31. The last item is sold at the t value at which the function is first equal to 100%. Thus the last item is sold at $t = 240$ or August 29.

   (b) Looking at the graph at $t = 121$ we see that roughly 65% of the inventory has been sold by May 1.

   (c) The percent of the inventory sold during May and June is the difference between the percent of the inventory sold on the first day of May and the percent of the inventory sold on the last day of June. Thus, the percent of the inventory sold during May and June is roughly 25%.

   (d) The percent of the inventory left after half a year is 100−(percent inventory sold after half year). Thus, roughly 10% of the inventory is left after half a year.

   (e) The items probably went on sale on day 100 and were on sale until day 120. Roughly from April 10 until April 30.

9. (a) The percentage of calls lasting from 1 to 2 minutes is given by the integral

$$\int_1^2 p(x)\,dx = \int_1^2 0.4e^{-0.4x}\,dx = e^{-0.4} - e^{-0.8} \approx 22.1\%.$$

   (b) A similar calculation (changing the limits of integration) gives the percentage of calls lasting 1 minute or less as

$$\int_0^1 p(x)\,dx = \int_0^1 0.4e^{-0.4x}\,dx = 1 - e^{-0.4} \approx 33.0\%.$$

   (c) The percentage of calls lasting 3 minutes or more is given by the improper integral

$$\int_3^\infty p(x)\,dx = \lim_{b \to \infty} \int_3^b 0.4e^{-0.4x}\,dx = \lim_{b \to \infty}(e^{-1.2} - e^{-0.4b}) = e^{-1.2} \approx 30.1\%.$$

   (d) The cumulative distribution function is the integral of the probability density; thus,

$$C(h) = \int_0^h p(x)\,dx = \int_0^h 0.4e^{-0.4x}\,dx = 1 - e^{-0.4h}.$$

13. (a) Most of the earth's surface is below sea level. Much of the earth's surface is either around 3 miles below sea level or exactly at sea level. It appears that essentially all of the surface is between 4 miles below sea level and 2 miles above sea level. Very little of the surface is around 1 mile below sea level.

    (b) The fraction below sea level corresponds to the area under the curve from −4 to 0 divided by the total area under the curve. This appears to be about $\frac{3}{4}$.

Answers for Section 6.6

1. The median daily catch is the amount of fish such that half the time a boat will bring back more fish and half the time a boat will bring back less fish. Thus the area under the curve and to the left of the median must be 0.5. There are 25 squares under the curve so the median occurs at 12.5 squares of area. Now

$$\int_2^5 p(x)dx = 11.0$$

and

$$\int_5^6 p(x)dx = 5.5$$

so the median occurs at a little over 5 tons. To find the value a for which

$$\int_5^a p(t)dt = 1.5$$

We note that we can approximate the integral with

$$\int_5^a p(t)dt = 5.5(a-5)$$

Thus we get

$$1.5 = 5.5(a-5)$$

and

$$a = 5 + \frac{1.5 \text{ tons}}{5.5} = 5.27 \text{ tons.}$$

5. We know that the mean is given by

$$\int_{-\infty}^{\infty} tp(t)dt.$$

Thus we get

$$\begin{aligned}\text{Mean} &= \int_0^4 tp(t)dt \\ &= \int_0^4 (-0.0375t^3 + 0.225t^2)dt \\ &= 0.075t^3 - 0.009375t^4 \Big|_0^4 \\ &= 2.4\end{aligned}$$

Thus the mean is 2.4 weeks.

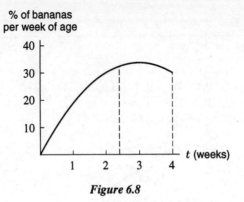

Figure 6.8

Figure 6.8 verifies that $t = 2.4$ is in fact the mean.

9. The median is value is the value T such that

$$\int_{-\infty}^{T} p(x)dx = 0.5$$

Thus we get

$$0.5 = \int_{-\infty}^{T} p(x)dx$$
$$= \int_{0}^{T} 0.122 e^{(-0.122x)} dx$$

Plugging in different values for T we get $T \approx 5.68$. Thus the median occurs at 5.68 seconds We know that the mean is

$$\int_{-\infty}^{\infty} xp(x)dx.$$

Thus we get

$$\text{Mean} = \int_{-\infty}^{\infty} xp(x)dx$$
$$= \int_{0}^{40} x(0.122 E^{-0.122x}) dx$$

Calculating the integral we get that the mean occurs at roughly 7.803 seconds.
The median tells us that fifty percent of the time there will be a time gap of less than 5.68 seconds between cars, and fifty percent of the time there will be a time gap of more than 5.68 seconds between cars.
The mean tells us that over a given interval of time, the average time gap between cars is 7.83 seconds.

13. (a) *Figure 6.9*

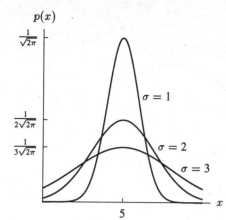

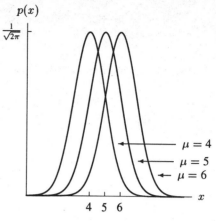

Figure 6.10

(b) Recall that the mean is the "balancing point." In other words, if the area under the curve was made of cardboard, we'd expect it to balance at the mean. All of the graphs are symmetric across the line $x = \mu$, so μ is the "balancing point" and hence the mean. As the graphs also show, increasing σ flattens out the graph, in effect lessening the concentration of the data near the mean. Thus, the smaller the σ value, the more data is clustered around the mean.

REVIEW PROBLEMS FOR CHAPTER SIX

1. Since $f(x)$ is positive along the interval from 0 to 6 the area is simply

$$\int_0^6 (x^2 + 2)dx = 84.$$

5.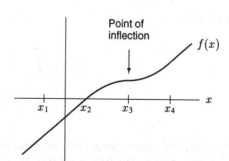

9. We approximate the area of the playing field by using Riemann sums. From the data provided,

$$\text{LEFT}(10) = \text{RIGHT}(10) = \text{TRAP}(10) = 89{,}000 \text{ square feet.}$$

Thus approximately

$$\frac{89{,}000 \text{ sq. ft.}}{200 \text{ sq. ft./lb.}} = 445 \text{ lbs. of fertilizer}$$

should be necessary.

13.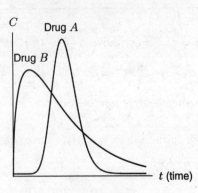

Figure 6.11

17. (a) The area under the graph of the height density function $p(x)$ is concentrated in two humps centered at 0.5 m and 1.1 m. The plants can therefore be separated into two groups, those with heights in the range 0.3 m to 0.7 m, corresponding to the first hump, and those with heights in the range 0.9 m to 1.3 m, corresponding to the second hump. This grouping of the grasses according to height is probably close to the species grouping. Since the second hump contains more area than the first, there are more plants of the tall grass species in the meadow.

(b) As do all cumulative distribution functions, the cumulative distribution function $P(x)$ of grass heights rises from 0 to 1 as x increases. Most of this rise is achieved in two spurts, the first as x goes from 0.3 m to 0.7 m, and the second as x goes from 0.9 m to 1.3 m. The plants can therefore be separated into two groups, those with heights in the range 0.3 m to 0.7 m, corresponding to the first spurt, and those with heights in the range 0.9 m to 1.3 m, corresponding to the second spurt. This grouping of the grasses according to height is the same as the grouping we made in part (a), and is probably close to the species grouping.

(c) The fraction of grasses with height less than 0.7 m equals $P(0.7) = 0.25 = 25\%$. The remaining 75% are the tall grasses.

CHAPTER 7

Answers for Section 7.1

1. Since $y = t^4$, the derivative is $\frac{dy}{dt} = 4t^3$. From the left side of the differential equation, we get

$$t\frac{dy}{dt} = t(4t^3) = 4t^4.$$

   Considering the right side, we find

$$4y = 4t^4.$$

   Since the substitution $y = t^4$ gives a true equality in the differential equation, $y = t^4$ is in fact a solution.

5. If $P = P_0 e^t$, then

$$\frac{dP}{dt} = \frac{d}{dt}(P_0 e^t) = P_0 e^t = P.$$

9. Yes. To see why, we solve the equation $13x\frac{dy}{dx} = y$ with $y = x^n$. To solve the equation, we first evaluate $\frac{dy}{dx} = \frac{d}{dx}(x^n) = nx^{n-1}$. The equation becomes

$$13x(nx^{n-1}) = x^n$$

   But $13x(nx^{n-1}) = 13n(x \cdot x^{n-1}) = 13nx^n$, so we have

$$13n(x^n) = x^n$$

   For $x \neq 0$, we divide through by x^n to get $13n = 1$, so $n = \frac{1}{13}$. Thus, $y = x^{\frac{1}{13}}$ is a solution.

13. We can check the equations on the right as possible solutions to the differential equations on the left.

    (a) (II) If $y = 3x$ then we get that $\frac{dy}{dx} = 3$. Plugging this into our differential equation we get

$$\frac{dy}{dx} = \frac{y}{x}$$
$$3 = \frac{3x}{x}$$
$$3 = 3.$$

    Thus we see that $y = 3x$ is a solution for the differential equation.

    (V) If $y = x$ then we get that $\frac{dy}{dx} = 1$. Substituting $y = 3e^x$ and $y' = 3e^x$ into the differential equation we get

$$\frac{dy}{dx} = \frac{y}{x}$$
$$1 = \frac{x}{x}$$
$$1 = 1.$$

    Thus $y = x$ is a second solution for the differential equation.

(b) (I) If $y = x^3$ then we get that $\frac{dy}{dx} = 3x^2$. Substituting $y = e^{3x}$ and $y' = 3e^{3x}$ into the differential equation we get

$$\frac{dy}{dx} = 3\frac{y}{x}$$
$$3x^2 = 3\frac{x^3}{x}$$
$$= 3x^2$$

Thus, $y = x^3$ solves this differential equation.

(c) Even if we try all the equations on the right, none satisfies this differential equation. To see why, notice that if $\frac{dy}{dx} = 3x$ then y is a function whose derivative is $3x$. Or in other words y is the anti-derivative of $3x$. Thus $y = \frac{3}{2}x^2 + k$ where k is any constant. Thus none of the functions are solutions.

(d) (IV) If $y = 3e^x$ then we get that $\frac{dy}{dx} = 3e^x$. Plugging this into our differential equation we get

$$\frac{dy}{dx} = y$$
$$3e^x = 3e^x$$

Thus we see that $y = 3e^x$ is a solution for the differential equation.

(e) (III) If $y = e^{3x}$ then we get that $\frac{dy}{dx} = 3e^{3x}$. Plugging this into our differential equation we get

$$\frac{dy}{dx} = 3y$$
$$3e^{3x} = 3e^{3x}$$

Thus we see that $y = e^{3x}$ is a solution for the differential equation.

Answers for Section 7.2

1. (a)

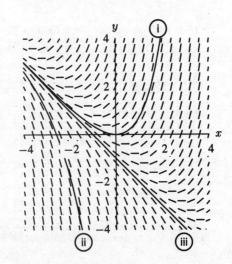

(b) The solution through $(-1, 0)$ appears to be linear, so its equation is $y = -x - 1$.

(c) If $y = -x - 1$, then $y' = -1$ and $x + y = x + (-x - 1) = -1$, so this checks as a solution.

5. When $a = 1$ and $b = 2$, the Gompertz equation is $y' = -y \ln(y/2) = y \ln(2/y) = y(\ln 2 - \ln y)$. This differential equation is similar to the differential equation $y' = y(2 - y)$ in certain ways. For example, in both equations y' is positive for $0 < y < 2$ and negative for $y > 2$. Also, for y values close to 2, $(\ln 2 - \ln y)$ and $(2 - y)$ are both close to 0, so $y(\ln 2 - \ln y)$ and $y(2 - y)$ are approximately equal to zero. Thus around $y = 2$ the slope fields look almost the same. This happens again around $y = 0$, since around $y = 0$ both $y(2 - y)$ and $y(\ln 2 - \ln y)$ go to 0. (Note that $\lim_{y \to 0^+} (y \ln y) = 0$.) For y values close to 1 the slope fields look similar since the local linearization of $\ln y$ near $y = 1$ is $y - 1$; hence, near $y = 1$, $y(\ln 2 - \ln y) \approx y(\ln 2 - (y - 1)) \approx y(1.69 - y) \approx y(2 - y)$. Finally, for $y > 2$, $\ln y$ grows much slower than y, so the slope field for $y' = y(\ln 2 - \ln y)$ is less steep, negatively, than for $y' = y(2 - y)$.

9. As $x \to \infty$, $y \to \infty$, no matter what the starting point.

13. (a) We know that when $x = 0$, $\frac{dy}{dx} = e^{x^2} = e^0 = 1$.

   We know that when $x = 1$, $\frac{dy}{dx} = e^{x^2} = e^1 = e$.

   We know that when $x = 2$, $\frac{dy}{dx} = e^{x^2} = e^4 \approx 54.6$.

   Thus the appropriate slope field is (IV).

(b) We know that when $x = 0$, $\frac{dy}{dx} = e^{-2x^2} = e^0 = 1$.

   We know that when $x = 1$, $\frac{dy}{dx} = e^{-2x^2} = e^{-2} \approx 0.14$.

   We know that when $x = 2$, $\frac{dy}{dx} = e^{-2x^2} = e^{-8} \approx 0.000335$.

   Thus the appropriate slope field is (I).

(c) We know that when $x = 0$, $\frac{dy}{dx} = e^{-x^2/2} = e^0 = 1$.

   We know that when $x = 1$, $\frac{dy}{dx} = e^{-x^2/2} = e^{-1/2} \approx 0.61$.

   We know that when $x = 2$, $\frac{dy}{dx} = e^{-x^2/2} = e^{-2} \approx 0.14$.

   Thus the appropriate slope field is (III).

(d) We know that when $x = 0$, $\frac{dy}{dx} = e^{-0.5x} \cos x = e^0 \cos 0 = 1$.

   We know that when $x = 1$, $\frac{dy}{dx} = e^{-0.5x} \cos x = e^{-1/2} \cos 1 \approx 0.33$.

   We know that when $x = 2$, $\frac{dy}{dx} = e^{-0.5x} \cos x = e^{-1} \cos 2 \approx -0.153$.

   Thus the appropriate slope field is (V).

(e) We know that when $x = 0$, $\frac{dy}{dx} = \frac{1}{(1 + 0.5 \cos x)^2} \approx 0.44$.

   We know that when $x = 1$, $\frac{dy}{dx} = \frac{1}{(1 + 0.5 \cos x)^2} \approx 0.61$.

   We know that when $x = 2$, $\frac{dy}{dx} = \frac{1}{(1 + 0.5 \cos x)^2} \approx 1.59$.

Thus the appropriate slope field is (II).

(f) We know that when $x = 0$, $\dfrac{dy}{dx} = -e^{-x^2} = -1$.

We know that when $x = 1$, $\dfrac{dy}{dx} = -e^{-x^2} \approx -0.37$.

We know that when $x = 2$, $\dfrac{dy}{dx} = -e^{-x^2} \approx -0.14$.

Thus the appropriate slope field is (VI).

Answers for Section 7.3

1. The equation given is in the form
$$\frac{dP}{dt} = kP.$$
Thus we know that the general solution to this equation will be
$$P = Ce^{kt}.$$
And in our case, with $k = 5$ we get
$$P = Ce^{0.02t}.$$

5. Rewriting we get
$$\frac{dy}{dx} = -\frac{1}{3}y.$$
We know that the general solution to an equation in the form
$$\frac{dy}{dx} = ky$$
is
$$y = Ce^{kx}.$$
Thus in our case we get
$$y = Ce^{-\frac{1}{3}x}.$$
We are told that $y(0) = 10$ so we get
$$y(x) = Ce^{-\frac{1}{3}x}$$
$$y(0) = 10 = Ce^0$$
$$C = 10$$
Thus we get
$$y = 10e^{-\frac{1}{3}x}.$$

9. In the equation $\frac{dy}{dt} = t^2$ we know the antiderivative of t^2 with respect to y. Anti-differentiating both sides we get the general solution
$$y = \frac{1}{3}t^3 + C$$
which has neither exponential growth nor exponential decay.

13. (a) = (I), (b) = (IV), (c) = (II) and (IV), (d) = (II) and (III).

17. (a) Since we are told that the rate at which the quantity of the drug decreases is proportional to the amount of the drug left in the body we know the differential equation modeling this situation is
$$\frac{dQ}{dt} = kQ.$$
Since we are told that the quantity of the drug is decreasing we know that $k < 0$.

(b) We know that the general solution to the differential equation
$$\frac{dQ}{dt} = kQ$$
is
$$Q = Ce^{kt}.$$

(c) We are told that the half life of the drug is 3.8 hours. This means that at $t = 3.8$ the amount of the drug in the body is half the amount that was in the body at $t = 0$, or in other words
$$0.5Q(0) = Q(3.8).$$
Solving this equation give
$$0.5Q(0) = Q(3.8)$$
$$0.5Ce^{k(0)} = Ce^{k(3.8)}$$
$$0.5C = Ce^{k(3.8)}$$
$$0.5 = e^{k(3.8)}$$
$$\ln(0.5) = k(3.8)$$
$$\frac{\ln(0.5)}{3.8} = k$$
$$k \approx -0.182.$$

(d) From part (c) we know that the formula for Q is
$$Q = Ce^{-0.182t}.$$
We are told that initially there are 10 mg of the drug in the body. Thus at $t = 0$ we get
$$10 = Ce^{-0.182(0)}$$
or simply
$$C = 10.$$
Thus our equation becomes
$$Q(t) = 10e^{-0.182t}.$$
Plugging in $t = 12$ we get
$$Q(t) = 10e^{-0.182t}$$
$$Q(12) = 10e^{-0.182(12)}$$
$$= 10e^{-2.184}$$
$$Q(12) \approx 1.126 \text{ mg}$$

21. Michigan:
$$\frac{dQ}{dt} = -\frac{r}{V}Q = -\frac{158}{4.9 \times 10^3}Q = -0.032Q$$

so
$$Q = Q_0 e^{-0.032t}$$

and
$$0.1Q_0 = Q_0 e^{-0.032t}$$

so
$$t = \frac{-\ln(0.1)}{0.032} \approx 71 \text{ years.}$$

Ontario:
$$\frac{dQ}{dt} = -\frac{r}{V}Q = \frac{-209}{1.6 \times 10^3}Q = -0.13Q$$

so
$$Q = Q_0 e^{-0.13t}$$

and
$$0.1Q_0 = Q_0 e^{-0.13t}$$

so
$$t = \frac{-\ln(0.1)}{0.13} \approx 18 \text{ years.}$$

Lake Michigan will take longer because it is larger (4900 km^3 compared to 1600 km^3) and water is flowing through it at a slower rate (158 km^3/year compared to 209 km^3/year).

Answers for Section 7.4

1. We know that the general solution to a differential equation of the form
$$\frac{dy}{dt} = k(y - A)$$

is
$$y = Ce^{kt} + A.$$

Thus in our case we get
$$y = Ce^{.5t} + 200.$$

We know that at $t = 0$ we have $y = 50$ so solving for C we get
$$y = Ce^{.5t} + 200$$
$$50 = Ce^{.5(0)} + 200$$
$$-150 = Ce^0$$
$$C = -150.$$

Thus we get
$$y = 200 - 150e^{.5t}.$$

5. We know that the general solution to a differential equation of the form
$$\frac{dP}{dt} = k(P - A)$$
is
$$P = Ce^{kt} + A.$$
Thus in our case we get
$$P = Ce^t - 4.$$
We know that at $t = 0$ we have $P = 100$ so solving for C we get
$$P = Ce^t - 4$$
$$100 = Ce^0 - 4$$
$$104 = Ce^0$$
$$C = 104.$$
Thus we get
$$P = 104e^t - 4.$$

9. (a) We know that the general solution to a differential equation of the form
$$\frac{dy}{dt} = k(y - A)$$
is
$$y = Ce^{kt} + A.$$
Factoring out a -1 on the left side we get
$$\frac{dy}{dt} = -(y - 100).$$
Thus in our case we get
$$y = Ce^{-t} + 100.$$
This is meaningful if $C \leq 0$, since one cannot know more than 100%.

(b)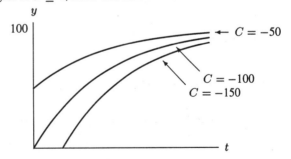

Figure 7.1

(c) Substituting $y = 0$ when $t = 0$ gives
$$0 = 100 - Ce^{-0}$$
so $C = 100$. Thus solution is
$$y = 100 - 100e^{-t}.$$

13. We find the temperature of the orange juice as a function of time. Newton's Law of Heating says that the rate of change of the temperature is proportional to the temperature difference. If S is the temperature of the juice, this gives us the equation

$$\frac{dS}{dt} = -k(S - 65) \text{ for some constant } k.$$

Note that k is positive, since $S = 40$ initially and S increases towards the temperature of the room. We know that the general solution to a differential equation of the form

$$\frac{dS}{dt} = k(S - A)$$

is

$$S = Ae^{kt} + A.$$

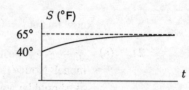

Thus in our case we get

$$S = Ae^{-kt} + 65.$$

Since at $t = 0$, $S = 40$, we have $40 = 65 + A$, so $A = -25$. Thus $S = 65 - 25e^{-kt}$ for some positive constant k.

17. (a) $\frac{dW}{dt} = \frac{1}{3500}(I - 20W)$

(b) We know that the general solution to a differential equation of the form

$$\frac{dW}{dt} = k(W - A)$$

is

$$W = Ce^{kt} + A.$$

Factoring out a -20 on the left side we get

$$\frac{dW}{dt} = \frac{-20}{3500}\left(W - \frac{-I}{-20}\right) = -\frac{2}{350}\left(W - \frac{1}{20}\right).$$

Thus in our case we get

$$W = Ce^{-\frac{2}{350}t} + \frac{I}{20}.$$

Let us call the person's initial weight W_0 at $t = 0$. Then $W_0 = \frac{I}{20} + Ae^0$, so $A = W_0 - \frac{I}{20}$. Thus

$$W = \frac{I}{20} + \left(W_0 - \frac{I}{20}\right)e^{-\frac{2}{350}t}.$$

(c) Using (b), we have $W = 150 + 15e^{-\frac{2}{350}t}$. This means that $W \to 150$ as $t \to \infty$.

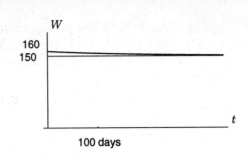

21. (a) $\frac{dp}{dt} = k(p - p_0)$, where k is the proportionality constant of the Evans Price Adjustment model. Notice that $k < 0$, since if $p > p_0$ then $\frac{dp}{dt}$ should be negative, and if $p < p_0$ then $\frac{dp}{dt}$ should be positive.

(b) We know that the general solution to a differential equation of the form

$$\frac{dp}{dt} = k(p - A)$$

is

$$p = Ce^{kt} + A.$$

Thus in our case we get

$$p = Ce^{kt} - p_0.$$

If I is the initial price, we get

$$p = Ce^{kt} - p_0$$
$$I = Ce^0 - p_0$$
$$C = I - p_0.$$

Thus we get

$$p = p_0 + (p - I)e^{kt}.$$

(c)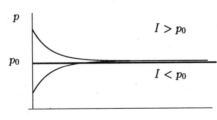

(d) As $t \to \infty, p \to p_0$. We see this in the solution in (b), since as $t \to \infty, e^{kt} \to 0$. (Remember $k < 0$!) In other words, as $t \to \infty$, p approaches the equilibrium price p_0.

25. (a) The quantity and the concentration both increase with time. As the concentration increases, the rate at which the drug is excreted also increases, and so the rate at which the drug builds up in the blood decreases; thus the graph of concentration against time is concave down. The concentration rises until the rate of excretion exactly balances the rate at which the drug is entering; at this concentration there is a horizontal asymptote. (See Figure 7.2.)

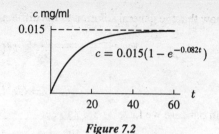

Figure 7.2

(b) Let's start by writing a differential equation for the quantity, $Q(t)$.

$$\begin{pmatrix} \text{Rate quantity} \\ \text{of drug changes} \end{pmatrix} = (\text{Rate in}) - (\text{Rate out})$$

$$\frac{dQ}{dt} = 43.2 - 0.082Q$$

where Q is measured in mg. We want an equation for concentration $c(t) = \frac{Q(t)}{v}$, where $c(t)$ is measured in mg/ml with $v = 35,000$ ml.

$$\frac{1}{v}\frac{dQ}{dt} = \frac{43.2}{v} - 0.082\frac{Q}{v},$$

giving

$$\frac{dc}{dt} = \frac{43.2}{35,000} - 0.082c.$$

(c) We know that the general solution to a differential equation of the form

$$\frac{dc}{dt} = k(c - B)$$

is

$$c = Ae^{kt} + B.$$

Factoring out a -0.082 on the left side we get

$$\frac{dc}{dt} = -0.082(c - 0.015).$$

Thus in our case we get

$$c = Ae^{-0.082t} + 0.015.$$

Since $c = 0$ when $t = 0$, we have $A = -0.015$, so

$$c = 0.015 - 0.015e^{-0.082t} = 0.015(1 - e^{-0.082t}).$$

Thus $c \to 0.015$ mg/ml as $t \to \infty$.

REVIEW PROBLEMS FOR CHAPTER SEVEN

1. Integrating both sides we get

$$y = \frac{5}{2}t^2 + C,$$

where C is a constant.

5. We know that the general solution to the differential equation
$$\frac{dy}{dx} = k(y - A)$$
is
$$y = Ce^{kx} + A.$$
Thus in our case we factor out 0.2 to get
$$\frac{dy}{dx} = 0.2\left(y - \frac{8}{0.2}\right) = 0.2(y - 40).$$
Thus the general solution to our differential equation is
$$y = Ce^{0.2x} + 40,$$
where C is some constant.

9. We know that the general solution to the differential equation
$$\frac{dP}{dt} = k(P - A)$$
is
$$P = Ce^{kt} + A.$$
Thus in our case we factor out 0.08 to get
$$\frac{dP}{dt} = 0.08\left(P - \frac{50}{0.08}\right) = 0.08(P - 625).$$
Thus the general solution to our differential equation is
$$P = Ce^{0.08t} + 625.$$
Solving for C with $P(0) = 10$ we get
$$P(t) = Ce^{0.08t} + 625$$
$$10 = Ce^0 + 625$$
$$C = -615$$
Thus the solution is
$$P = 625 - 615e^{0.08t}.$$
The graph of this function is shown in Figure 7.3

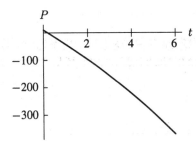

Figure 7.3

13. (a) k is positive because if $T > A$, then the body will lose heat so that its temperature falls to A. Thus, $\frac{dT}{dt}$ should be negative, so k should be positive. Similarly, if $T < A$ then $\frac{dT}{dt}$ should be positive, so k again should be positive.

(b) The units for $\frac{dT}{dt}$ is $\frac{\text{degrees}}{\text{time}}$. Since the units for $T - A$ are degrees, the units for k are $(\text{time})^{-1}$ or $\frac{1}{\text{time}}$. Thus, if we change from days to hours, k would look different – for example, $\frac{1}{1\,\text{day}} = \left(\frac{1}{24}\right)\left(\frac{1}{1\,\text{hour}}\right)$. In words, daily $k = 24$(hourly k). Similarly, the numerical value of k in $\frac{1}{1\,\text{hour}}$ is 60 times bigger than that of k expressed in $\frac{1}{1\,\text{min}}$.

(c) Everything else being equal, coffee will cool faster in a thin china cup than in styrofoam. Thus, for a given temperature difference, $\frac{dT}{dt}$ should be larger in magnitude for china, and therefore k should be larger.

(d) We have $\frac{dT}{dt} = -0.14(T - 70)$, and $T = 170$ when $t = 0$
$\frac{dT}{T-70} = -0.14\,dt$. Solving this, we have $T = 70 + Ae^{-0.14t}$
The initial condition gives us that $170 = 70 + A$, so $A = 100$. Therefore the solution to the differential equation is $T = 70 + 100e^{-0.14t}$.
To find the amount of time we must wait for the coffee to cool to 120 degrees or less, we put $T = 120$ into the solution and solve for t.

$$120 = 70 + 100e^{-0.14t} \text{ so } t \approx 5 \text{ minutes.}$$

To find how soon we must drink the coffee before it cools to less than 90 degrees, we put $T = 90$ into the solution and solve for t.

$$90 = 70 + 100e^{-0.14t} \text{ so } t \approx 11.5 \text{ minutes.}$$

Therefore our fussiness requires that we not drink our coffee until it has cooled for 5 minutes and then we must finish it before it has cooled for more than 11 minutes. Of course as we drink it, it will cool faster as we increase the surface area to volume ratio!

CHAPTER 8

Answers for Section 8.1

1. (a) 80-90°F (b) 60-72°F (c) 60-100°F

5. (a) Beef consumption by households making $20,000/year

    TABLE 8.1

p	3.00	3.50	4.00	4.50
$f(20, p)$	2.65	2.59	2.51	2.43

    i.e. $f(20, p)$ is a decreasing function of p.

    (b) Beef consumption by households making $100,000/year

    TABLE 8.2

p	3.00	3.50	4.00	4.50
$f(100, p)$	5.79	5.77	5.60	5.53

    i.e. $f(100, p)$ is also a decreasing function of p.

    (c) Beef consumption by households when the price of beef is $3.00/lb

    TABLE 8.3

I	20	40	60	80	100
$f(I, 3.00)$	2.65	4.14	5.11	5.35	5.79

    i.e. $f(I, 3.00)$ is an increasing function of I.

    (d) Beef consumption by households when the price of beef is $4.00/lb

    TABLE 8.4

I	20	40	60	80	100
$f(I, 4.00)$	2.51	3.94	4.97	5.19	5.60

    i.e. $f(I, 4.00)$ is also an increasing function of I.

9.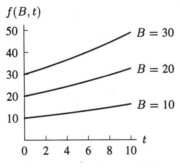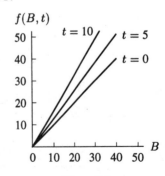

   Figure 8.1   Figure 8.2

The bigger the value fixed for B, the bigger the amount $f(B,t)$ is. No matter what $B > 0$ is, $f(B,t)$ is an increasing function of t. The bigger the fixed value of t, the bigger the influence of the increase in B on the increase of $f(B,t)$ (slopes of the lines in Figure 8.2.

13.

TABLE 8.5

°F\mph	5	10	15	20	25
20	16	3	−5	−10	−15

TABLE 8.6

°F\mph	5	10	15	20	25
0	−5	−22	−31	−39	−44

17. Asking if f is an increasing or decreasing function of p is the same as asking how does f vary as we vary p, when we hold a fixed. Intuitively, we know that as we increase the price p, total sales of the product will go down. Thus, f is a decreasing function of p. Similarly, if we increase a, the amount spent on advertising, we can expect f to increase and therefore f is an increasing function of a.

21. Q is a decreasing function of c and an increasing function of t. This is because when the price of coffee rises, consumers drink less. When the tea price rises, some consumers switch from tea to coffee and the demand for coffee increases.

25. The function $f(x,0) = \cos 0 \sin x = \sin x$ gives the displacement of each point of the string when time is held fixed at $t = 0$. The function $f(x,1) = \cos 1 \sin x = 0.54 \sin x$ gives the displacement of each point of the string at time $t = 1$. Graphing $f(x,0)$ and $f(x,1)$ gives in each case an arch of the sine curve, the first with amplitude 1 and the second with amplitude 0.54. For each different fixed value of t, we get a different snapshot of the string, each one a sine curve with amplitude given by the value of $\cos t$. The result looks like the sequence of snapshots shown in Figure 8.3.

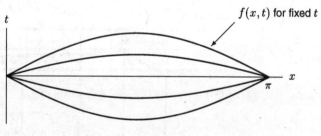

Figure 8.3

Answers for Section 8.2

1. B is closest to the yz-plane, since it has the smallest x-coordinate in absolute value. B lies on the xz-plane, since its y-coordinate is 0. B is farthest from the xy-plane, since it has the largest z-coordinate in absolute value.

5. Which way are you facing when you start? Since you start out behind the yz-plane (your x-coordinate is negative) and you are facing the yz-plane, you are facing in the direction of the positive x-axis. When you walk forward 3 units, you are walking 3 units in the positive x-direction, so your new x-coordinate is $-2 + 3 = 1$. You are still facing in the positive x-direction, so when you turn to your right, you are turning to walk in the negative y-direction. You move 1 unit in this direction, so your new y-coordinate is $-1 - 1 = -2$. Finally, when you move down 4 units, you are subtracting 4 from your z-coordinate, so your new z-coordinate is $2 - 4 = -2$. Thus the coordinates of your final position are

$$(-2 + 3, -1 - 1, 2 - 4) = (1, -2, -2).$$

9. The graph is all points with $y = 4$ and $z = 2$, i.e., a line parallel to the x-axis and passing through the points $(0, 4, 2); (2, 4, 2); (4, 4, 2)$ etc.

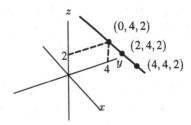

Figure 8.4

13. The x-coordinate of the the center equals $-1 + \frac{5}{2} = 1.5$. The y-coordinate of the center equals $-2 + \frac{5}{2} = 0.5$. The z-coordinate of the center equals $2 - \frac{5}{2} = -0.5$.

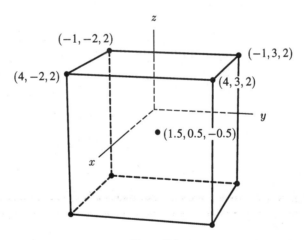

Figure 8.5

104

17.

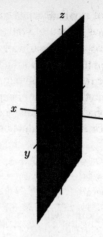

Figure 8.6

Answers for Section 8.3

1. (a) Decreases:

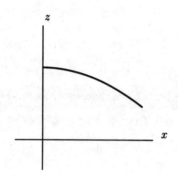

Figure 8.7

(b) Increases:

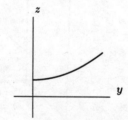

Figure 8.8

5. (a) The value of z only depends on the distance from the point (x, y) to the origin. Therefore the graph has a circular symmetry around the z-axis. There are two such graphs among those depicted in Figure 8.35: I and V. The one corresponding to $z = \frac{1}{x^2+y^2}$ is I since the function blows up as (x, y) gets close to $(0, 0)$.

(b) For similar reasons as in part (a), the graph is circularly symmetric about the z-axis, hence the corresponding one must be V.

(c) The graph has to be a plane, hence IV.

(d) The function is independent of x, hence the corresponding graph can only be II. Notice that the cross-sections of this graph parallel to the yz-plane are parabolas, which is a confirmation of the result.

(e) The graph of the given function is depicted in III. The picture shows clearly the cross-sections parallel to the zx-plane. They have the familiar shape of the cubic curves $z = x^3 -$ constant. It is a little bit harder to see the sine curves $z = $ constant $- \sin y$ corresponding to $x = $ constant.

9. (a)

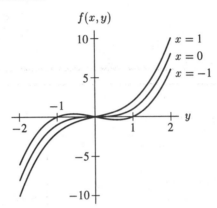

Figure 8.9

(b)

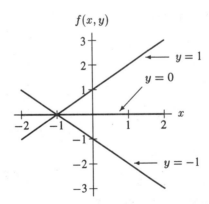

Figure 8.10

Answers for Section 8.4

1. The contour where $f(x, y) = x + y = c$ or $y = -x + c$ is a graph of the straight line of slope -1 as shown in Figure 8.11. Note that we have plotted the region where $-4 < c < 4$. The contours are evenly spaced.

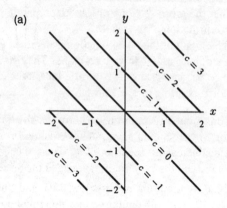

Figure 8.11

5. The contour where $f(x, y) = 2x - y = c$ is a graph of the straight line $y = 2x - c$ of slope 2 as shown in Figure 8.12. Note that we have plotted the region where $-6 < c < 6$. The contours are evenly spaced.

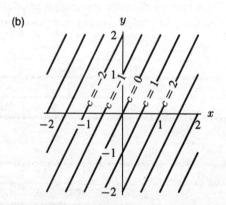

Figure 8.12

9. The contour where $f(x, y) = xy = c$ is a graph of the rectangular-hyperbola $y = c/x$ as shown in Figure 8.13. Note that we have plotted the region where $-6 < c < 6$. The contours become more closely packed as we move further from the origin.

(c)

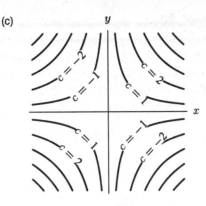

Figure 8.13

13. Using our economic intuition, we know that the total sales of a product should be an increasing function of the amount spent on advertising. From the graph, Q is a decreasing function of x and an increasing function of y. Thus, the y-axis corresponds to the amount spent on advertising and the x-axis corresponds to the price of the product.

17. (a) Finding the point representing 13% and $6000 on the graph, we note it lies between the 120 and 140 contours. Interpolating between these values we estimate the monthly payment to be $\approx$ $137. See Figure 8.61.
    (b) Since the interest rate has dropped, we will be able to borrow more money and still make a monthly payment of $137. To find out how much we can afford to borrow, we should find where 11% intersects with the $137 contour and read over to the loan amount to which these values correspond. Since the $137 contour is not shown, we must estimate its position from the $120 and $140 contours. When we do this, we find that we can borrow an amount of money that is more than $6000 but less than $6500. So we can borrow about $250 more without increasing the monthly payment.
    (c) The entries in the table will be the price at which each interest rate intersects the 137 contour. Using the $137 contour from (b) we make the table as follows.

TABLE 8.7 *Amount borrowed at a monthly payment of $137.*

Interest Rate (%)	0	1	2	3	4	5	6	7
Loan Amount ($)	8200	8000	7800	7600	7400	7200	7200	6800

TABLE 8.8

Interest rate (%)	8	9	10	11	12	13	14	15
Loan Amount ($)	6650	6500	6350	6250	6100	6000	5900	5800

21. One possible answer follows.
    (a) If there is a city at the center of the diagram, then the population is very dense at the center, but progressively less dense as you move into the suburbs, further from the city center. This scenario corresponds to diagram (I) or (II). We pick (I) because it has the highest density, as we would expect in a city.

(b) If the center of the diagram is a lake, and is a very busy and thriving center, where lake front property is considered the most desirable, then the most dense area will be at lakeside, and decrease as you move further from the lake in the center, as in diagram (I) or (II). We pick (II) because we expect the population density at lake front to be less than that in the center of a city.

(c) If the center of the diagram is a power plant and if the plant is not in a densely populated area, where people can and will choose not to live anywhere near it, the population density will then be very low nearby, increasing slightly further from the plant, as in diagram (III).

In an alternative solution, if the lake were in the middle of nowhere, the entire area would be very sparsely populated, and there would be slightly fewer people living on the actual lake shore, as in diagram (III).

25. (a) East-west cross-section along the line $N = 30°$

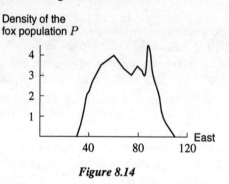

Figure 8.14

(b) East-west cross-section along the line $N = 60°$

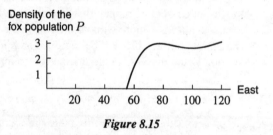

Figure 8.15

(c) North-south cross-section at along the line $E = 40°$

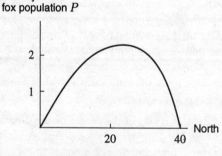

Figure 8.16

(d) North-south cross-section along the line $E = 80°$

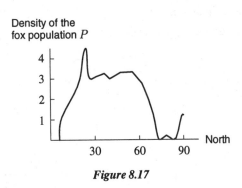

Figure 8.17

29. (a) Observe that the values in this table are symmetric about the origin (i.e. $(x, y) = (-x, -y)$), which means that the contours cannot be the lines shown in (I) or (IV). Also observe that as you move away from the origin, whose contour value is 0, the z values on the contours increase. Thus, this table corresponds to diagram (II).
    (b) This table is similarly symmetric about the origin, however the contour values are decreasing as you move away from the origin, which is shown in diagram (III).
    (c) This table shows that for each value of x, every value of y gives the same contour value, suggesting a straight vertical line at each x value, as in diagram (IV).
    (d) This table also shows lines, however these are horizontal since for each value of y any x value gives the same contour value. Thus, this table represents diagram (I).

33. (a) (III) (K)
    (b) (V) (G)
    (c) (IV) (J)
    (d) (I) (H)
    (e) (II) (F)

Answers for Section 8.5

1. Since $z = 2x + 3y + C$, $\Delta z = 2\Delta x + 3\Delta y$.
   Now $\Delta x = 0.5$, $\Delta y = -0.2$, so $\Delta z = 2 \cdot 0.5 + 3 \cdot (-0.2) = 0.4$.
   Since $C = z - 2x - 3y = 2 - 2 \cdot 5 - 3 \cdot 7 = -29$,
   therefore $z = 2 \cdot 4.9 + 3 \cdot 7.2 - 29 = 2.4$ when $x = 4.9$ and $y = 7.2$.

5. No. Not all of the columns and rows are linear, so the function is not linear.

9.

TABLE 8.9

$x \backslash y$	0.0	1.0
0.0	−1.0	1.0
2.0	3.0	5.0

13. In the diagram the contours correspond to values of the function that are 200 units apart, i.e., there are contours for 1000, 1200, 1400, etc. Approximately moving one unit in the y direction we cross one contour; i.e., a change of 1 in y changes the function by 200, so the y slope is 200. Similarly, a move of 100 in the x direction crosses three contour lines and changes the function by 600; so the x slope is 6. Hence $f(x,y) = c + 6x + 200y$. We see from the diagram that $f(50, 0) = 400$, so $c = 100$. Therefore the function is $f(x,y) = 100 + 6x + 200y$.

17.

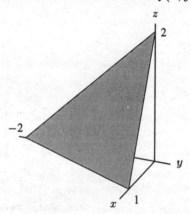

Figure 8.18

Answers for Section 8.6

1. (a)
$$Q = 900L^{\frac{1}{2}}K^{\frac{2}{3}}$$
$$= 900 \cdot 70^{\frac{1}{2}} \cdot 50^{\frac{2}{3}}$$
$$= 102{,}197$$

(b) If L and K are doubled, plug into the equation and see
$$\tilde{Q} = 900(2L)^{\frac{1}{2}}(2K)^{\frac{2}{3}}$$
$$= 900L^{\frac{1}{2}}K^{\frac{2}{3}} \cdot (2)^{\frac{1}{2}}(2)^{\frac{2}{3}}$$
$$= 2^{\frac{7}{6}} \cdot Q$$

so Q will increase by a factor of 2.2. Plugging in $L = 140$ and $K = 100$, $\tilde{Q} = 229,425 = 2^{\frac{7}{6}} \cdot Q$

5.

Function	Graph	Statement
$F(L,K) = L^{0.25}K^{0.25}$	(II)	(E)
$F(L,K) = L^{0.5}K^{0.5}$	(I)	(D)
$F(L,K) = L^{0.75}K^{0.75}$	(III)	(G)

We get the information above because for $F(L,K) = bL^\alpha K^\beta$, we have:
constant returns to scale for $\alpha + \beta = 1$,
increasing returns to scale for $\alpha + \beta > 1$, and
decreasing returns to scale for $\alpha + \beta < 1$.

REVIEW PROBLEMS FOR CHAPTER EIGHT

1.

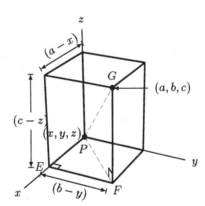

Figure 8.19

The z-coordinate of a point is the distance from the point to the xy-plane and the y-coordinate is the distance from the xz-plane, and the x-coordinate is the distance from the yz-plane. To see this, imagine a box a units deep, b units long, and c units high, with one corner at the origin, as in Figure 8.19. By following along the edges in the figure, you can see that the corner of the box diagonally opposite the origin is the point (a, b, c), since you get there by going a units along the x-axis, b units parallel to the y-axis, and c units parallel to the z-axis. Picture this in terms of the room you are in. On the other hand, since the box has depth a you can see that the point (a, b, c) is a units from the yz-plane; since it has length b you can see that the point is b units from the xz-plane; and since it has height c you can see that the point is c units from the xy-plane.

5. Could not be true. If the origin is on the level curve $z = 1$, then $z = f(0,0) = 1 \neq -1$. So $(0,0)$ cannot be on both $z = 1$ and $z = -1$.

9. True. For every point (x, y), compute the value $z = e^{-(x^2+y^2)}$ at that point. The level curve obtained by getting z equal to that value goes through the point (x, y).

13. Contours are lines of the form $y = \frac{3x}{5} + \frac{1-c}{5}$ as shown in Figure 8.20. Note that for the regions of x and y given, the c values range from $-15 < c < 17$.

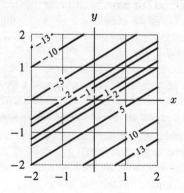

Figure 8.20

17. The point $x = 10, t = 5$ is between the contours $H = 70$ and $H = 75$, a little closer to the former. So we estimate $H(10, 5) \approx 72$, i.e., it is about $72°$F. Five minutes later we are at the point $x = 10, t = 10$, which is just above the contour $H = 75$, so we estimate that it has warmed up to $76°$F by then.

21.

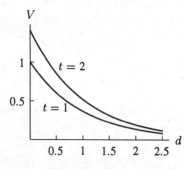

Figure 8.21

As you get farther from the volcano, the fallout declines exponentially. At the start of the eruption, there is no fallout; as time passes, the fallout increases. This answer models the behavior of an actual volcano, as you would expect the fallout to increase with time but diminish rapidly with distance.

25. To read off the cross-sections of f with t fixed, we choose a t value and move horizontally across the diagram looking at the values on the contours. For $t = 0$, as we move from the left at $x = 0$ to the right at $x = \pi$, we cross contours of $0.25, 0.50, 0.75$ and reach a maximum at $x = \pi/2$, and then decrease back to 0. That is because if time is fixed at $t = 0$, then $f(x, 0)$ is the displacement of the string at that time: no displacement at $x = 0$ and $x = \pi$ and greatest displacement at $x = \pi/2$. For cross-sections with t fixed at larger values, as we move along

a horizontal line, we cross fewer contours and reach a smaller maximum value: the string is becoming less curved. At time $t = \pi/2$, the string is straight so we see a value of 0 all the way across the diagram, namely a contour with value 0. For $t = \pi$, the string has vibrated to the other side and the displacements are negative as we read across the diagram reaching a minimum at $x = \pi/2$.

The cross-sections of f with x fixed are read vertically. At $x = 0$ and $x = \pi$, we see vertical contours of value 0 because the end points of the string have 0 displacement no matter what time it is. The cross-section for $x = \pi/2$ is found by moving vertically up the diagram at $x = \pi/2$. As we expect, the contour values are largest at $t = 0$, zero at $t = \pi/2$, and a minimum at $t = \pi$.

Notice that the spacing of the contours is also important. For example, for the $t = 0$ cross-section, contours are most closely spaced at the end points at $x = 0$ and $x = \pi$ and most spread out at $x = \pi/2$. That is because the shape of the string at time $t = 0$ is a sine curve, which is steepest at the end points and relatively flat in the middle. Thus, the contour diagram shows the steepest terrain at the end points and flattest terrain in the middle.

CHAPTER 9

Answers for Section 9.1

1. If h is small, then
$$f_x(3,2) \approx \frac{f(3+h,2) - f(3,2)}{h}.$$

With $h = 0.01$, we find
$$f_x(3,2) \approx \frac{f(3.01,2) - f(3,2)}{0.01} = \frac{\frac{3.01^2}{(2+1)} - \frac{3^2}{(2+1)}}{0.01} = 2.00333.$$

With $h = 0.0001$, we get
$$f_x(3,2) \approx \frac{f(3.0001,2) - f(3,2)}{0.0001} = \frac{\frac{3.0001^2}{(2+1)} - \frac{3^2}{(2+1)}}{0.0001} = 2.0000333.$$

Since the difference quotient seems to be approaching 2 as h gets smaller, we conclude
$$f_x(3,2) \approx 2.$$

To estimate $f_y(3,2)$, we use
$$f_y(3,2) \approx \frac{f(3,2+h) - f(3,2)}{h}.$$

With $h = 0.01$, we get
$$f_y(3,2) \approx \frac{f(3,2.01) - f(3,2)}{0.01} = \frac{\frac{3^2}{(2.01+1)} - \frac{3^2}{(2+1)}}{0.01} = -0.99668.$$

With $h = 0.0001$, we get
$$f_y(3,2) \approx \frac{f(3,2.0001) - f(3,2)}{0.0001} = \frac{\frac{3^2}{(2.0001+1)} - \frac{3^2}{(2+1)}}{0.0001} = -0.9999667.$$

Thus, it seems that the difference quotient is approaching -1, so we estimate
$$f_y(3,2) \approx -1.$$

5. (a) This means you must pay a mortgage payment of $1090.08/month if you have borrowed a total of $92000 at an interest rate of 14%, on a 30-year mortgage.
   (b) This means that the rate of change of the monthly payment with respect to the interest rate is $72.82; i.e., your monthly payment will go up by approximately $72.82 for each percentage point increase in the interest rate.
   (c) It should be *positive*, because the monthly payments will increase if the total amount borrowed is increased.
   (d) It should be *negative*, because as you increase the number of years in which to pay the mortgage, you should have to pay less each month.

9. (a) For points near the point $(0, 5, 3)$, moving in the positive x direction, the surface is sloping down and the function is decreasing. Thus, $f_x(0, 5) < 0$.

(b) Moving in the positive y direction near this point the surface slopes up as the function increases, so $f_y(0, 5) > 0$.

13. Since the average rate of change of the wind-chill factor is about -2.6 (drops by $2.6°$F) with every 1 mph increase in wind speed from 5 mph to 10 mph, when the temperature stays constant at $20°$F, we know that

$$f_w(5, 20) \approx -2.6$$

17. Estimating from the contour diagram, we have, for point A,

$$\frac{\partial n}{\partial x}(A) \approx \frac{3-2}{45-40} = \frac{1 \text{ fox}}{5 \text{ mi}}$$
$$\frac{\partial n}{\partial y}(A) \approx \frac{1-2}{36-30} = -\frac{1 \text{ fox}}{6 \text{ mi}}.$$

So, from point A the fox population increases (just barely) as we move eastward. The population decreases as we move north from A.

At point B,

$$\frac{\partial n}{\partial x}(B) \approx \frac{1\frac{1}{2}-2}{90-80} = -\frac{1/2 \text{ fox}}{10 \text{ mi}}$$
$$\frac{\partial n}{\partial y}(B) \approx \frac{1-2}{70-60} = -\frac{1 \text{ fox}}{10 \text{ mi}}.$$

So, fox population decreases in both directions east and north of B. Seems to decrease more as we move north of B.

At point C,

$$\frac{\partial n}{\partial x}(C) \approx \frac{4-3}{100-70} = \frac{1 \text{ fox}}{30 \text{ mi}}$$
$$\frac{\partial n}{\partial y}(C) \approx \frac{4-3}{45-35} = \frac{1 \text{ fox}}{10 \text{ mi}}.$$

So, the fox population increases as we move east and north of C. Greater increase as we move north of C.

21. The partial derivative $H_w(t, w)$ can be approximated by

$$H_w(10, 0.1) \approx \frac{H(10, 0.1 + h) - H(10, 0.1)}{h} \quad \text{for small } h.$$

We choose $h = 0.1$ because we can read off a value for $H(10, 0.1 + 0.1) = H(10, 0.2)$ from the graph. If we take $H(10, 0.2) = 240$, we get the approximation

$$H_w(10, 0.1) \approx \frac{H(10, 0.1 + 0.1) - H(10, 0.1)}{0.1} = \frac{240 - 110}{0.1} = 1300.$$

In practical terms, we have found that for fog at $10°$ C containing 0.1 g/m^3 of water, an increase in the water content of the fog will increase the heat requirement for dissipating the fog at the rate given by $H_w(10, 0.1)$. Wetter fog is harder to dissipate. Other values of $H_w(t, w)$ in Table 9.1 are computed using the formula

$$H_w(t, w) \approx \frac{H(t, w + 0.1) - H(t, w)}{0.1},$$

where we have used Table ?? to evaluate H.

TABLE 9.1 *Table of values of $H_w(t, w)$ (in cal/gm)*

t (°C)	w (gm/m^3)		
	0.1	0.2	0.3
10	1300	900	1200
20	800	800	900
30	800	700	800

Answers for Section 9.2

1. (a) The difference quotient for approximating $f_u(u, v)$ is given by

   $$f_u(u, v) \approx \frac{f(u + h, v) - f(u, v)}{h}.$$

   Putting in $(u, v) = (1, 3)$ and $h = 0.001$, the difference quotient is

   $$\begin{aligned} f_u(1, 3) &\approx \frac{2(1.001)^2 \cdot 3 - 2(1)^2(3)}{0.001} \\ &\approx \frac{6.0120 - 6}{0.001} = \frac{0.0120}{0.001} \\ &\approx 12.00 \end{aligned}$$

   (b) Using the derivative formulas

   $$f_u = \frac{\partial f}{\partial u} = 4uv$$

   so

   $$f_u(1, 3) = 4(1)(3) = 12$$

   . We see that the approximation in part (a) was reasonable!

5. $f_y = \frac{1}{2}(x^2 + y^2)^{-\frac{1}{2}} \cdot (2y) = \frac{y}{\sqrt{x^2 + y^2}}$

9. $f_x = \frac{200x}{y}$, $f_y = -\frac{100x^2}{y^2}$

13. $\frac{\partial A}{\partial h} = \frac{1}{2}(a + b)$

17. $F_v = \frac{2mv}{r}$

21. $\frac{\partial}{\partial t}(v_0 t + \frac{1}{2}at^2) = v_0 + \frac{1}{2} \cdot 2at = v_0 + at$

25. $\frac{dB}{dt} = Pre^{rt}$, it is rate of growth of money as time passes. $\frac{dB}{dP} = e^{rt}$, it tells you for a given rate r, how much more money you would have if you had deposited one more dollar.

29. (a) $Q_K = 18.75 K^{-0.25} L^{0.25}$, $Q_L = 6.25 K^{0.75} L^{-0.75}$.
    (b) When $K = 60$ and $L = 100$,

$$Q = 25 \cdot 60^{0.75} \cdot 100^{0.25} = 1704.33$$
$$Q_K = 18.75 \cdot 60^{-0.25} 100^{0.25} = 21.3041$$
$$Q_L = 6.25 \cdot 60^{0.75} 100^{-0.75} = 4.2608$$

(c) Q is actual quantity being produced. Q_K is how much more could be produced if you increased K by one unit. Q_L is how much more could be produced if you increased L by 1.

33. $f_{xx} = 0$, $f_{xy} = -2/y^2$, $f_{yy} = 4x/y^3$, $f_{yx} = -2/y^2$
37. $V_{rr} = 2\pi h$, $V_{hh} = 0$, $V_{rh} = V_{hr} = 2\pi r$
41. $P_{rr} = 100t^2 e^{rt}$, $P_{tt} = 100r^2 e^{rt}$, $P_{rt} = P_{tr} = 100rte^{rt}$

Answers for Section 9.3

1. Mississippi lies entirely within a region designated as "80s" so we expect both the high and low daily temperatures within the state to be in the 80s. The South-Western most corner of the state is close to a region designated as "90s" so, we would expect the temperature here to be in the high 80s, say 87-88. The northern most portion of the state is relatively central to the "80s" region. We might expect the temperature there to be between 83-87.

   Alabama also lies completely within a region designated as "80s" so both the high and low daily temperatures within the state are in the 80s. The south-eastern tip of the state is close to a "90s" region so we would expect the temperature here to be $\approx$ 88-89 degrees. The northern most part of the state is relatively central to the "80s" region so the temperature there is $\approx$ 83-87 degrees.

   Pennsylvania is also in the "80s" region, but it is touched by the boundary line between the "80s" and a "70s" region. Thus we expect the low daily temperature to occur there and be about 70 degrees. The state is also touched by a boundary line which contains a "90s" region so the high will occur there and be 89-90 degrees.

   New York is split by a boundary between an "80s" and a "70s" region, the northern portion of the state is apt to be about 74-76 while the southern portion is likely to be in the low 80s, maybe 81-84 or so.

   California contains many different zones. The northern coastal areas will probably have the state daily low at 65-68, although without another contour on that side, it is difficult to judge how quickly the temperature is dropping off to the west. The tip of Southern California is in a 100s region, so there we expect the state daily high to be 100-101.

   Arizona will have a low around 85-87 in the northwest corner and a high in the 100s, perhaps 102-107 in its southern regions.

   Massachusetts will probably have a high around 81-84 and a low at 70.

5. To maximize $z = x^2 + y^2$, it suffices to maximize x^2 and y^2. We can maximize both of these at the same time by taking the point $(1, 1)$, where $z = 2$. It occurs on the boundary of the square. (Note: We also have maxima at the points $(-1, -1), (-1, 1)$ and $(1, -1)$ which are on the boundary of the square.)

   To minimize $z = x^2 + y^2$, we choose the point $(0, 0)$, where $z = 0$. It does not occur on the boundary of the square.

9. At a critical point $f_x = 2x + 4 = 0$ and $f_y = 2y = 0$, so it is the point $(-2, 0)$.

13. For Problem 8, at the critical point $f_{xx} \cdot f_{yy} - f_{xy}^2 = 2 \cdot 2 - 0 = 4 > 0$, and $f_{xx} > 0$, so it is a local minimum point.

    For Problem 9, at the critical point $f_{xx} \cdot f_{yy} - f_{xy}^2 = 2 \cdot 2 - 0 = 4 > 0$, and $f_{xx} > 0$, so it is a local minimum point.

    For Problem 10, at the critical point $f_{xx} \cdot f_{yy} - f_{xy}^2 = 2 \cdot 0 - 1 = -1 < 0$, so it is a saddle point.

    For Problem 11, at the critical point $f_{xx} \cdot f_{yy} - f_{xy}^2 = 0 \cdot 12 - 9 = -9 < 0$, so it is a saddle point.

17. At the origin $f(0, 0) = 0$. Since $x^6 \geq 0$ and $y^6 \geq 0$, the point $(0, 0)$ is a local (and global) minimum. The second derivative test does not tell you anything since $D = 0$.

21. (a) $(1, 3)$ is a critical point. Since $f_{xx} > 0$ and the discriminant
    $$D = f_{xx} f_{yy} - f_{xy}^2 = f_{xx} f_{yy} - 0^2 = f_{xx} f_{yy} > 0,$$
    the point $(1, 3)$ is a minimum.

    (b)

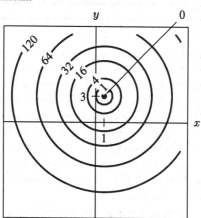

Figure 9.1

Answers for Section 9.4

1. We first express the revenue R in terms of the prices p_1 and p_2:
   $$\begin{aligned} R(p_1, p_2) &= p_1 q_1 + p_2 q_2 \\ &= p_1(517 - 3.5p_1 + 0.8p_2) + p_2(770 - 4.4p_2 + 1.4p_1) \\ &= 517p_1 - 3.5p_1^2 + 770p_2 - 4.4p_2^2 + 2.2p_1 p_2. \end{aligned}$$

At a local maximum we have $\frac{\partial R}{\partial p_1} = 0$ and $\frac{\partial R}{\partial p_2} = 0$, and so:

$$\frac{\partial R}{\partial p_1} = 517 - 7p_1 + 2.2p_2 = 0,$$

$$\frac{\partial R}{\partial p_2} = 770 - 8.8p_2 + 2.2p_1 = 0.$$

Solving these equations, we find that

$$p_1 = 110 \quad \text{and} \quad p_2 = 115.$$

To see whether or not we have a found a local maximum, we compute the second-order partial derivatives:

$$\frac{\partial^2 R}{\partial p_1^2} = -7, \quad \frac{\partial^2 R}{\partial p_2^2} = -8.8, \quad \frac{\partial^2 R}{\partial p_1 \partial p_2} = 2.2.$$

Therefore,

$$D = \frac{\partial^2 R}{\partial p_1^2}\frac{\partial^2 R}{\partial p_2^2} - \frac{\partial^2 R}{\partial p_1 \partial p_2} = (-7)(-8.8) - (2.2)^2 = 56.76,$$

and so we have found a local maximum point. The graph of $R(p_1, p_2)$ has the shape of an upside down paraboloid and so $(110, 115)$ is a global maximum point.

5. Let $P(K, L)$ be the profit obtained using K units of capital and L units of labor. The cost of production is given by

$$C(K, L) = kK + \ell L,$$

and the revenue function is given by

$$R(K, L) = pQ = pAK^a L^b.$$

Hence, the profit is

$$P = R - C = pAK^a L^b - (kK + \ell L).$$

In order to find local maxima of P, we calculate the partial derivatives and see where they are zero. We have:

$$\frac{\partial P}{\partial K} = apAK^{a-1}L^b - k,$$

$$\frac{\partial P}{\partial L} = bpAK^a L^{b-1} - \ell.$$

The critical points of the function $P(K, L)$ are solutions (K, L) of the simultaneous equations:

$$\frac{k}{a} = pAK^{a-1}L^b,$$

$$\frac{\ell}{b} = pAK^a L^{b-1}.$$

Multiplying the first equation by K and the second by L, we get

$$\frac{kK}{a} = \frac{\ell L}{b},$$

and so

$$K = \frac{\ell a}{kb}L.$$

Substituting for K in the equation $k/a = pAK^{a-1}L^b$, we get:

$$\frac{k}{a} = pA\left(\frac{\ell a}{kb}\right)^{a-1} L^{a-1}L^b.$$

We must therefore have

$$L^{1-a-b} = pA\left(\frac{a}{k}\right)^a \left(\frac{\ell}{b}\right)^{a-1}.$$

Hence, if $a + b \neq 1$,

$$L = \left[pA\left(\frac{a}{k}\right)^a \left(\frac{\ell}{b}\right)^{(a-1)}\right]^{(a+b-1)},$$

and

$$K = \frac{\ell a}{kb}L = \frac{\ell a}{kb}\left[pA\left(\frac{a}{k}\right)^a \left(\frac{\ell}{b}\right)^{(a-1)}\right]^{(a+b-1)}.$$

To see if this is really a local maximum, we apply the second derivative test. We have:

$$\frac{\partial^2 P}{\partial K^2} = a(a-1)pAK^{a-2}L^b,$$

$$\frac{\partial^2 P}{\partial L^2} = b(b-1)pAK^a L^{b-2},$$

$$\frac{\partial^2 P}{\partial K \partial L} = abpAK^{a-1}L^{b-1}.$$

Hence,

$$D = \frac{\partial^2 P}{\partial K^2}\frac{\partial^2 P}{\partial L^2} - \left(\frac{\partial^2 P}{\partial K \partial L}\right)^2$$
$$= ab(a-1)(b-1)p^2A^2K^{2a-2}L^{2b-2} - a^2b^2p^2A^2K^{2a-2}L^{2b-2}$$
$$= ab((a-1)(b-1) - ab)p^2A^2K^{2a-2}L^{2b-2}$$
$$= ab(1-a-b)p^2A^2K^{2a-2}L^{2b-2}.$$

Now a, b, p, A, K, and L are positive numbers. So, the sign of this last expression is determined by the sign of $1 - a - b$. We assumed that $a + b < 1$, so $D > 0$, and as $0 < a < 1$, then $\partial^2 P/\partial K^2 < 0$ and so we have a unique local maximum. Since $a + b < 1$, if we move along any line outward from the origin in the KL-plane towards infinity, the profit $P(K, L)$ will eventually become negative and so we have found the global maximum.

(b) In the case $a + b < 1$, we had decreasing returns to scale. That is, if the amount of capital and labor used is multiplied by a constant $\lambda > 0$, we get less than λ times the production. Now suppose $a + b \geq 1$. If we multiply K and L by λ for some $\lambda > 0$, then

$$Q(\lambda K, \lambda L) = A(\lambda K)^a(\lambda L)^b = \lambda^{a+b}Q(K, L).$$

We also see that

$$C(\lambda K, \lambda L) = \lambda C(K, L)$$

and so if $a + b = 1$, we have

$$P(\lambda K, \lambda L) = \lambda P(K, L).$$

Thus, if $\lambda = 2$, so we are doubling the inputs K and L, then the profit P is doubled and hence there can be no maximum profit. If $a + b > 1$, we have increasing returns to scale and there can again be no maximum profit: doubling the inputs will more than double the profit.

9. (a) Points which are directly above or below each other share the same x coordinate, therefore, the point on the least squares line which is directly above or below the point in question will have x coordinate x_i and from the formula for the least squares line, it will have y coordinate $b + mx_i$.

(b) The general distance formula in two dimensions is $d = \sqrt{(x_2 - x_1)^2 + (y_2 - y_1)^2}$, so $d^2 = (x_2 - x_1)^2 + (y_2 - y_1)^2$. Since the x coordinates are identical for the two points in question, the first term in the square root is zero. This yields $d^2 = (y_i - (b + mx_i))^2$.

(c) In both cases we use the chain rule and our knowledge of summations to show the relationship.

$$\frac{\partial f}{\partial b} = \frac{\partial}{\partial b}(\sum_{i=1}^{n}(y_i - (b + mx_i))^2) = \sum_{i=1}^{n}\frac{\partial}{\partial b}(y_i - (b + mx_i))^2$$

$$= \sum_{i=1}^{n} 2(y_i - (b + mx_i)) \cdot \frac{\partial}{\partial b}(y_i - (b + mx_i))$$

$$= \sum_{i=1}^{n} 2(y_i - (b + mx_i)) \cdot (-1)$$

$$= -2\sum_{i=1}^{n}(y_i - (b + mx_i))$$

$$\frac{\partial f}{\partial m} = \frac{\partial}{\partial m}(\sum_{i=1}^{n}(y_i - (b + mx_i))^2) = \sum_{i=1}^{n}\frac{\partial}{\partial m}(y_i - (b + mx_i))^2$$

$$= \sum_{i=1}^{n} 2(y_i - (b + mx_i)) \cdot \frac{\partial}{\partial m}(y_i - (b + mx_i))$$

$$= \sum_{i=1}^{n} 2(y_i - (b + mx_i)) \cdot (-x_i)$$

$$= -2\sum_{i=1}^{n}(y_i - (b + mx_i)) \cdot x_i$$

(d) We can separate $\frac{\partial f}{\partial b}$ into three sums as shown:

$$\frac{\partial f}{\partial b} = -2\left(\sum_{i=1}^{n} y_i - b\sum_{i=1}^{n} 1 - m\sum_{i=1}^{n} x_i\right)$$

Similarly we can separate $\frac{\partial f}{\partial m}$ after multiplying through by x_i:

$$\frac{\partial f}{\partial m} = -2\left(\sum_{i=1}^{n} y_i x_i - b\sum_{i=1}^{n} x_i - m\sum_{i=1}^{n} x_i^2\right)$$

Setting $\frac{\partial f}{\partial b}$ and $\frac{\partial f}{\partial m}$ equal to zero we have:

$$0 = \left(\sum_{i=1}^{n} y_i\right) - bn - \left(m\sum_{i=1}^{n} x_i\right)$$

$$0 = \left(\sum_{i=1}^{n} y_i x_i\right) - \left(b \sum_{i=1}^{n} x_i\right) - m\left(\sum_{i=1}^{n} x_i^2\right)$$

(e) Solving this pair of simultaneous equations we get the result:

$$b = \left(\sum_{i=1}^{n} x_i^2 \sum_{i=1}^{n} y_i - \sum_{i=1}^{n} x_i \sum_{i=1}^{n} y_i x_i\right) \bigg/ \left(n \sum_{i=1}^{n} x_i^2 - \left(\sum_{i=1}^{n} x_i\right)^2\right)$$

$$m = \left(n \sum_{i=1}^{n} y_i x_i - \sum_{i=1}^{n} x_i \sum_{i=1}^{n} y_i\right) \bigg/ \left(n \sum_{i=1}^{n} x_i^2 - \left(\sum_{i=1}^{n} x_i\right)^2\right)$$

(f) Applying the formula to the given data set results gives the least squares line $y = -(1/3) + x$, in perfect agreement with the example.

13. (a) The principle indicates that the relationship between A and N is governed by a power function, that is $N = kA^p$. This is clear because as one increases by a factor of a, the other increases by a factor of b. If the relationship was linear, an increase by a factor a in one would result in an increase by a factor of a in the other. If the relationship was exponential, a linear increase a in one would result in an exponential increase by a factor a in the other (a and b are arbitrary factors). This case is covered by none of those possibilities and furthermore, we can see that in the case of a power function such as $y = x^p$, an increase by a factor of a in x will be seen as an increase by a factor of a^p in y. According to this rule of thumb, $a = 10$ follows $a^p = 2$. So $p = \log 2 \approx 0.301$. Thus $N \approx kA^{0.301}$.

(b)
$$\ln N = \ln(kA^p)$$
$$\ln N = \ln k + p \ln A$$
$$\ln N \approx \ln k + 0.301 \ln A$$

Thus, $\ln N$ is a linear function of $\ln A$.

(c) Table 9.2 contains the natural logarithms of the data:

TABLE 9.2 $\ln N$ and $\ln A$

Island	$\ln A$	$\ln N$
Redonda	0.00	1.61
Saba	2.08	2.20
Montserrat	4.32	2.71
Puerto Rico	8.15	4.32
Jamaica	8.35	4.25
Hispaniola (Haiti and Dominican Rep.)	10.29	4.87
Cuba	10.70	4.83

Using a least squares fit we find the line:

$$\ln N = 1.538 + 0.320 \ln A$$

This yields the power function:

$$N = e^{1.538} A^{0.320} = 4.655 A^{0.320}$$

Since 0.320 is pretty close to $\log 2 (\approx 0.301)$, the answer does agree with the biological rule.

Answers for Section 9.5

1. Our objective function is $f(x,y) = x+y$ and our equation of constraint is $g(x,y) = x^2+y^2 = 1$. To optimize $f(x,y)$ with Lagrange multipliers, we solve

$$f_x = \lambda g_x$$
$$f_y = \lambda g_y$$
$$g(x,y) = 1$$

Since

$$f_1 = 1 \qquad f_y = 1$$
$$g_x = 2x \qquad g_y = 2y$$

we have

$$1 = \lambda \cdot 2x$$
$$1 = \lambda \cdot 2y$$

Solving for λ gives

$$\lambda = \frac{1}{2x} = \frac{1}{2y}.$$

Which tells us $x = y$. Going back to our equation of constraint, we use the substitution $x = y$ to solve for y

$$g(y,y) = y^2 + y^2 = 1$$
$$2y^2 = 1$$
$$y^2 = \frac{1}{2}$$
$$y = \pm\sqrt{\frac{1}{2}} = \pm\frac{\sqrt{2}}{2}.$$

Since $x = y$, our critical points are $(\frac{\sqrt{2}}{2}, \frac{\sqrt{2}}{2})$ and $(-\frac{\sqrt{2}}{2}, -\frac{\sqrt{2}}{2})$. Evaluating f at these points we find the maximum value is $f(\frac{\sqrt{2}}{2}, \frac{\sqrt{2}}{2}) = \sqrt{2}$ and the minimum value is $f(-\frac{\sqrt{2}}{2}, -\frac{\sqrt{2}}{2}) = -\sqrt{2}$.

5. The objective function is $f(x,y) = x^2 + y^2$ and the equation of constraint is $g(x,y) = x^4 + y^4 = 2$. To optimize $f(x,y)$ with Lagrange multipliers, we solve

$$f_x = \lambda g_x$$
$$f_y = \lambda g_y$$
$$g(x,y) = 2$$

we have

$$f_x = 2x \qquad f_y = 2y$$
$$g_x = 4x^3 \qquad g_y = 4y^3$$

This tells us that

$$2x = \lambda 4x^3$$
$$2y = \lambda 4y^3.$$

Now if $x = 0$, the first equation is true for any value of λ. In particular, we can choose λ which satisfies the second equation. Similarly, $y = 0$ is solution.

Assuming both $x \neq 0$ and $y \neq 0$, we can divide to solve for λ and find

$$\lambda = \frac{2x}{4x^3} = \frac{2y}{4y^3}$$
$$\frac{1}{2x^2} = \frac{1}{2y^2}$$
$$y^2 = x^2$$
$$y = \pm x.$$

Going back to our equation of constraint, we find

$$g(0, y) = 0^4 + y^4 = 2 \Rightarrow y = \pm\sqrt[4]{2}$$
$$g(x, 0) = x^4 + 0^4 = 2 \Rightarrow x = \pm\sqrt[4]{2}$$
$$g(x, \pm x) = x^4 + (\pm x)^4 = 2 \Rightarrow x = \pm 1.$$

Thus, the critical points are $(0, \pm\sqrt[4]{2})$, $(\pm\sqrt[4]{2}, 0)$, $(1, \pm 1)$ and $(-1, \pm 1)$. Evaluating f at these points we find

$$f(1, 1) = f(1, -1) = f(-1, 1) = f(-1, -1) = 2,$$
$$f(0, \sqrt[4]{2}) = f(0, -\sqrt[4]{2}) = f(\sqrt[4]{2}, 0) = f(-\sqrt[4]{2}, 0) = \sqrt{2}.$$

So the minimum value of $f(x, y)$ on $g(x, y) = 2$ is $\sqrt{2}$ and the maximum value is 2.

9. (a) To be producing the maximum quantity Q under the cost constraint given, the firm should be using K and L values given by $Q_K = \lambda C_K, Q_L = \lambda C_L$ and $C = 20K + 10L = 150$, so

$$\frac{\partial Q}{\partial K} = 0.6aK^{-0.4}L^{0.4} = 20\lambda$$
$$\frac{\partial Q}{\partial L} = 0.4aK^{0.6}L^{-0.6} = 10\lambda$$

Hence $\dfrac{0.6aK^{-0.4}L^{0.4}}{0.4aK^{0.6}L^{-0.6}} = 1.5\dfrac{L}{K} = \dfrac{20\lambda}{10\lambda} = 2$, so $L = \dfrac{4}{3}K$. Substituting in $20K + 10L = 150$, we obtain $20K + 10\left(\dfrac{4}{3}\right)K = 150$. Then $K = \dfrac{9}{2}$ and $L = 6$, so capital should be reduced by $\dfrac{1}{2}$ unit, and labor should be increased by 1 unit.

(b) $\dfrac{\text{New production}}{\text{Old production}} = \dfrac{a4.5^{0.6}6^{0.4}}{a5^{0.6}5^{0.4}} \approx 1.01$, so tell the board of directors, "Reducing the quantity of capital by $1/2$ unit and increasing the quantity of labor by 1 unit will increase production by 1% while holding costs to \$150."

13. We want to minimize the function $h(x,y)$ subject to the constraint that

$$g(x,y) = x^2 + y^2 = (2,000)^2 = 4,000,000.$$

Using the method of Lagrange multipliers, we obtain the following system of equations:

$$-\frac{10x + 4y}{24,000} = 2\lambda x,$$
$$-\frac{4x + 4y}{24,000} = 2\lambda y,$$
$$x^2 + y^2 = 4,000,000.$$

If $x = 0$, the first equation would imply that $y = 0$. But the point $(0,0)$ does not satisfy the third equation and so we can assume $x \neq 0$. Similarly, if $y = 0$, the second equation would imply that $x = 0$, and so we can also assume that $y \neq 0$. We can now eliminate λ from the first and second equations, to give

$$2y^2 + 3xy - 2x^2 = (2y - x)(y + 2x) = 0,$$

and so the climber either moves along the line $x = 2y$ or $y = 2x$. We must now choose one of these lines and a direction along that line which will lead to the point of minimum height on the circle. To do this we find the points of intersection of these lines with the circle $x^2 + y^2 = 4,000,000$, compute the corresponding heights, and then select the minimum point. If $x = 2y$, the third equation gives

$$5y^2 = (2,000)^2,$$

so that $y = \pm 2,000/\sqrt{5} \approx \pm 894.4272$ and $x = \pm 1788.8542$. If $y = 2x$, we find that $x = \pm 894.4272$ and $y = \pm 1788.8542$. The corresponding heights are

$$h(\pm 894.4272, \pm 1788.8542) = 9,300,$$
$$h(\pm 1788.8542, \pm 894.4272) = 9,000.$$

Therefore, she should travel along the line $y = 2x$, in either of the two possible directions.

17. We wish to minimize the objective function

$$C(x, y, z) = 20x + 10y + 5z$$

subject to the budget constraint

$$Q(x, y, z) = 20x^{1/2} y^{1/4} z^{2/5} = 1,200.$$

Therefore, we solve the equations

$$C_x = \lambda Q_x$$
$$C_y = \lambda Q_y$$
$$C_z = \lambda Q_z$$
$$Q = 1,200$$

We have

$$20 = 10\lambda x^{-1/2}y^{1/4}z^{2/5} \quad \text{or} \quad \lambda = 2x^{1/2}y^{-1/4}z^{-2/5},$$
$$10 = 5\lambda x^{1/2}y^{-3/4}z^{2/5}, \quad \text{or} \quad \lambda = 2x^{-1/2}y^{3/4}z^{-2/5},$$
$$5 = 8\lambda x^{1/2}y^{1/4}z^{-3/5}, \quad \text{or} \quad \lambda = 0.625x^{-1/2}y^{-1/4}z^{3/5},$$
$$20x^{1/2}y^{1/4}z^{2/5} = 1,200.$$

The first and second equations imply that

$$x = y,$$

while the second and third equations imply that

$$3.2y = z.$$

Substituting for x and z in the fourth equation gives

$$y \approx 23.47,$$

and so

$$x \approx 23.47 \quad \text{and} \quad z \approx 75.1.$$

REVIEW PROBLEMS FOR CHAPTER NINE

1. $f_x = 2x + y$, $f_y = 2y + x$.

5. $\dfrac{\partial P}{\partial K} = 7K^{-0.3}L^{0.3}$, $\dfrac{\partial P}{\partial L} = 3K^{0.7}L^{-0.7}$,

9. The partial derivative, $\partial Q/\partial b$ is the rate of change of the quantity of beef purchased with respect to the price of beef, when the price of chicken stays constant. If the price of beef increases and the price of chicken stays the same, we expect consumers to buy less beef and more chicken. Thus when b increases, we expect Q to decrease, so $\partial Q/\partial b < 0$.

   On the other hand, $\partial Q/\partial c$ is the rate of change of the quantity of beef purchased with respect to the price of chicken, when the price of beef stays constant. An increase in the price of chicken is likely to cause consumers to buy less chicken and more beef. Thus when c increases, we expect Q to increase, so $\partial Q/\partial c > 0$.

13. (a) $\dfrac{\partial q_1}{\partial x}$ is the rate of change of the quantity of the first brand sold as its price increases. Since this brand competes with another brand, an increase in the price of the first brand should result in a decrease in the quantity sold of this same brand. Thus $\dfrac{\partial q_1}{\partial x} < 0$. Similarly, $\dfrac{\partial q_2}{\partial y} < 0$.

    (b) Again we take into consideration the competition between the two brands. If the second brand were to increase its price, then more of the first brand should sell. Thus $\dfrac{\partial q_1}{\partial y} > 0$. Similarly, $\dfrac{\partial q_2}{\partial x} > 0$.

17. Note that the x-axis and the y-axis are not in the domain of f.

(a) Since $x \neq 0$ and $y \neq 0$, by setting $f_x = 0$ and $f_y = 0$ we get

$$f_x = 1 - \frac{1}{x^2} = 0 \text{ when } x = \pm 1$$

$$f_y = 1 - \frac{4}{y^2} = 0 \text{ when } y = \pm 2$$

So the critical points are $(1,2), (-1,2), (1,-2), (-1,-2)$. Since $f_{xx} = 2/x^3$ and $f_{yy} = 8/y^3$ and $f_{xy} = 0$, the discriminant is

$$D(x,y) = f_{xx}f_{yy} - f_{xy}^2 = \left(\frac{2}{x^3}\right)\left(\frac{8}{y^3}\right) - 0^2 = \frac{16}{(xy)^3}.$$

Since $D < 0$ at the points $(-1, 2)$ and $(1, -2)$, these points are saddle points. Since $D > 0$ at $(1, 2)$ and $(-1, -2)$ and $f_{xx}(1, 2) > 0$ and $f_{xx}(-1, -2) < 0$, the point $(1, 2)$ is a local minimum and the point $(-1, -2)$ is a local maximum.

(b) No global maximum or minimum, since $f(x, y)$ increases without bound if x and y increase in the first quadrant; $f(x, y)$ decreases without bound if x and y decrease in the third quadrant.

21. Since $z_x = 8x - y = 0$ and $z_y = -x + 8y = 0$ then $x = 0, y = 0$. So $(0, 0)$ is the only critical point within the disc.
On the boundary, using Lagrange multipliers, let

$$G = x^2 + y^2 - 2 = 0,$$
$$z_x = \lambda G_x$$
$$z_y = \lambda G_y$$

and

$$G_x = 2x$$
$$G_y = 2y$$

So $8x - y = \lambda 2x$ and $-x + 8y = \lambda 2y$. Thus $\dfrac{8x - y}{2x} = \lambda = \dfrac{-x + 8y}{2y}$. Solving we get, $x^2 = y^2$ or $x = \pm y$.
From $G = 0$, we get $2x^2 = 2$, so $x^2 = 1$, so $x = \pm 1$.
The critical points are thus $(1, 1), (-1, 1), (1, -1), (-1, -1), (0, 0)$. Substituting into z, we get, $z(1, 1) = 7, z(-1, -1) = 7, z(1, -1) = 9, z(-1, 1) = 9, z(0, 0) = 0$. Thus $(-1, 1)$ and $(1, -1)$ give the maxima over the disc and $(0, 0)$ gives the minimum.

APPENDIX

Answers for Appendix A

1. The graph is

    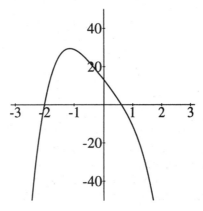

    (a) The range appears to be $y \leq 30$.
    (b) The function has two zeros.

5. The largest root is at about 2.5.

9. Using a graphing calculator, we see that when x is around 0.45, the graphs intersect.

13. (a) Only one real zero, at about $x = -1.15$.
    (b) Three real zeros: at $x = 1$, and at about $x = 1.41$ and $x = -1.41$.

17. (a) Since f is continuous, there must be one zero between $\theta = 1.4$ and $\theta = 1.6$, and another between $\theta = 1.6$ and $\theta = 1.8$. These are the only clear cases. We might also want to investigate the interval $0.6 \leq \theta \leq 0.8$ since $f(\theta)$ takes on values close to zero on at least part of this interval. Now, $\theta = 0.7$ is in this interval, and $f(0.7) = -0.01 < 0$, so f changes sign twice between $\theta = 0.6$ and $\theta = 0.8$ and hence has two zeros on this interval (assuming f is not *really* wiggly here, which it's not). There are a total of 4 zeros.

    (b) As an example, we find the zero of f between $\theta = 0.6$ and $\theta = 0.7$. $f(0.65)$ is positive; $f(0.66)$ is negative. So this zero is contained in $[0.65, 0.66]$. The other zeros are contained in the intervals $[0.72, 0.73]$, $[1.43, 1.44]$, and $[1.7, 1.71]$.

    (c) You've found all the zeros. A picture will confirm this; see Figure 0.1.

    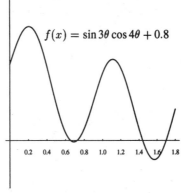

    Figure 0.1:

NOTES

NOTES

NOTES

NOTES

NOTES

NOTES

NOTES

NOTES

NOTES

NOTES